International Mathematical Olympiad Volume III

Anthem Science, Technology and Medicine

International Mathematical Olympiad Volume III

1991–2004

ISTVAN REIMAN

Anthem Press

Anthem Press
An imprint of Wimbledon Publishing Company
75-76 Blackfriars Road, London SE1 8HA
or
PO Box 9779, London SW19 7ZG
www.anthempress.com

This edition published by Anthem Press 2005.
First published by Typotex Ltd in Hungarian as
Nemzetközi Matematikal Diákolimpiák by I Reiman.

Translated by János Pataki, András Stipsitz & Csaba Szabó

British Library Cataloguing in Publication Data
A catalogue record for this book is available from the British Library.

Library of Congress Cataloguing in Publication Data
A catalogue record for this book has been requested.

1 3 5 7 9 10 8 6 4 2

ISBN 1 84331 203 4 (Hbk)
ISBN 1 84331 204 2 (Pbk)

Cover Illustration: Footprint Labs

Prelims typeset by Footprint Labs Ltd, London
www.footprintlabs.com

Printed in India

Preface

This three volume set contains the problems from the first forty-five IMO-s, from 1959 to 2004.

The chronicle of the IMO (International Mathematics Olympiad) starts with the initiative of the Romanian Mathematics and Physics Society: in July 1959 on the occasion of a celebration the Society invited high school students from the neighbouring countries to an international mathematical competition. The event proved to be such a success that the participants all agreed to go on with the enterprise. Ever since, this competition has taken place annually (except for 1980) and it has gradually transformed from the local contest of but a few countries into the most important and comprehensive international mathematical event for the young. Only seven nations were invited for the first IMO, while the number of participating countries was well beyond eighty for the last event; wherever mathematical education has reached a moderate level, sooner or later the country has turned up at the IMO.

The movement has had a significant impact on the mathematical education of several participating countries and also on the development of the gifted. The aim of a more proficient preparation for the IMO itself has launched the organization of national mathematical competitions in many countries involved. As the crucial component of successful participation, the preparation for the contest has enriched the publishing activity in several countries. Math-clubs have been formed on a large scale and periodicals have started. Even though the competition certainly brings up some pedagogical problems, if the educators regard the competitions not as ultimate aims, but as ways to introduce and endear pupils to mathematics, then their pedagogical benefit is undeniable.

The administration of the competitions has not changed that much; the larger scale has obviously necessitated certain modifications but the actual contest is more or less as it used to be. The participating countries are invited to delegate a group of up to six students who are attending high school at the year of the contest or had just finished their secondary school studies. Three problems are posed each day over two consecutive days and the students have to produce written solutions in their native tongue. There are two delegation leaders accompanying each team; one of their tasks is to provide an oral translation of their students' work into one of the official languages—by now this has been almost exclusively English—for a committee of mathematicians from the host country. Together with this group of coordinators they eventually settle the score the solutions are worth; the highest mark is seven points for each problem. The contestants are then ranked according to their total scores; the awarding of the prizes has been administered according to the following principle: half of the participants are given a prize: namely the proportion of the gold, silver and bronze medals is $1:2:3$ respectively.

The occasional professional problems are handled by the international jury formed by the leaders of the participating delegations; their most important and difficult task is to select the six problems for the actual contest, to formulate their official text and to prepare rough marking schemes for each of them. The organizers ask for proposals from the participating countries well in advance; in due course they produce a list of approximately twenty to twenty five problems from those suggested and the jury selects the final six from this supply.

There are almost two hundred problems in this three volume set and they provide a full image of the challenge the students had to cope with during these forty years. One cannot claim that every single one of them is a pearl of mathematics but their overwhelming majority is interesting and rewarding; together they more or less cover the usual syllabus-chapters of elementary mathematics. When selecting, the jury usually tries to choose from the intersection of the respective curricula of the participating countries; considering that there are more than eighty of them this is not an easy job, if not impossible. The reader might notice that there are no problems at all from the theory of probability, for example, and complex numbers hardly show up.

From the retrospect of more than forty years one can certainly conclude that the IMO movement has had a significant role in the history of the second half of twentieth century mathematics. There are quite a few highly ranked mathematicians who started their career at an IMO; even at this point, however, we have to emphasize, that an eventual fiasco at the IMO or any other mathematical contest whatsoever usually has no implications at all about the mathematical potential of a well prepared student.

A careful reader will certainly realize that quite a few problems in this book are in fact simplifications or particular cases of more profound mathematical results; apart from the intellectual satisfaction of actually solving these problems, the discovery of this mathematical background and the knowledge gained from it can be the ultimate benefits of a high level study of this book.

At the end of the book we included a Glossary of Theorems (and their proofs) we used in the book and we refer to them by their numbers enclosed in brackets, e. g. [6].

István Reiman

International Mathematical Olympiad

Problems

1991

1991/1. Given a triangle ABC, let I be the incentre. The internal bisectors of angles A, B and C meet the opposite sides at A', B', C' respectively. Prove that

(1) $$\frac{1}{4} < \frac{AI \cdot BI \cdot CI}{AA' \cdot BB' \cdot CC'} \leq \frac{8}{27}.$$

1991/2. Let $n > 6$ be an integer and let a_1, a_2, $\ldots$, a_k be all the positive integers less than n and relatively prime to n. If

$$a_2 - a_1 = a_3 - a_2 = \ldots = a_k - a_{k-1} > 0,$$

prove that n must be either a prime number or a power of 2.

1991/3. Let $S = \{1, 2, 3, \ldots, 280\}$. Find the smallest integer n such that each n-element subset of S contains five numbers which are pairwise relatively prime.

1991/4. Suppose G is a connected graph with k edges. Prove that it is possible to label the edges $1, 2, 3, \ldots, k$ in such a way that at each vertex which belongs to two or more edges, the greatest common divisor of the integers labelling those edges is 1.

[A graph is a set of points, called vertices, together with a set of edges joining certain pairs of distinct vertices. Each pair of points belongs to at most one edge. The graph is connected if for each pair of distinct vertices x, y there is some sequence of vertices $x = v_0, v_1, \ldots, v_m = y$, such that each pair v_i, v_{i+1} $(0 \leq i < m)$ is joined by an edge.]

1991/5. Let ABC be a triangle and X an interior point of ABC. Show that at least one of the angles XAB, XBC, PCA is less than or equal to $30°$.

1991/6. Given any real number $a > 1$ construct a bounded infinite sequence $x_0, x_1, \ldots$ such that

(1) $$|x_i - x_j| \cdot |i - j|^a \geq 1$$

for every pair of distinct i, j. [An infinite sequence $x_0, x_1, \ldots$ of real numbers is bounded if there is a constant C such that $|x_i| < C$ for all i.]

1992

1992/1. Find all integers a, b, c satisfying $1 < a < b < c$ such that $(a-1) \cdot (b-1) \cdot (c-1)$ is a divisor of $abc - 1$.

1992/2. Find all functions f defined on the set of all real numbers with real values, such that

(1) $$f\left(x^2 + f(y)\right) = y + (f(x))^2$$

for all x, y.

1992/3. Consider 9 points in space, no 4 coplanar. Each pair of points is joined by a line segment which is coloured either blue or red or left uncoloured. Find the smallest value of n such that whenever exactly n edges are coloured, the set of coloured edges necessarily contains a triangle all of whose edges have the same colour.

1992/4. L is tangent to the circle C and M is a point on L. Find the locus of all points P such that there exist points Q and R on L equidistant from M with C the incircle of the triangle PQR.

1992/5. Let S be a finite set of points in three-dimensional space. Let S_x, S_y, S_z be the sets consisting of the orthogonal projections of the points of S onto the yz-plane, xz-plane, xy-plane respectively. Prove that
$$|S|^2 \le |S_x||S_y||S_z|,$$
where $|A|$ denotes the number of points in the set A.

1992/6. For each positive integer n, $S(n)$ is defined as the greatest integer such that for every positive integer $k \le S(n)$, n^2 can be written as the sum of k positive squares.

(a) Prove that $S(n) \le n^2 - 14$ for each $n \ge 4$.

(b) Find an integer n such that $S(n) = n^2 - 14$.

(c) Prove that there are infinitely many integers n such that $S(n) = n^2 - 14$.

1993

1993/1. Let $f(x) = x^n + 5x^{n-1} + 3$, where $n > 1$ is an integer. Prove that $f(x)$ cannot be expressed as the product of two non-constant polynomials with integer coefficients.

1993/2. Let D be a point inside the acute-angled triangle ABC such that $\angle ABD = 90° + \angle ACB$ and $AC \cdot BD = AD \cdot BC$.

(a) Calculate the ratio $\dfrac{AB \cdot CD}{AC \cdot BD}$.

(b) Prove that the tangents at C to the circumcircles of ACD and BCD are perpendicular.

1993/3. On an infinite chessboard a game is played as follows. At the start n^2 pieces are arranged in an $n \times n$ block of adjoining squares, one piece on each square. A move in the game is a jump in a horizontal or vertical direction over an adjacent occupied square to an unoccupied square immediately beyond. The piece which has been jumped over is removed. Find those values of n for which the game can end with only one piece remaining on the board.

1993/4. For three points P, Q, R in the plane define $m(PQR)$ as the minimum length of the three altitudes of the triangle PQR (or zero if the points are collinear). Prove that for any points A, B, C, X

$$m(ABC) \leq m(ABX) + m(AXC) + m(XBC).$$

1993/5. Does there exist a function f from the positive integers to the positive integers such that $f(1) = 2$, $f(f(n)) = f(n) + n$ for all n, and $f(n) < f(n+1)$ for all n?

1993/6. There are $n > 1$ lamps L_0, L_1, ..., L_{n-1} in a circle. We use L_{n+k} to mean L_k. A lamp is at all times either on or off. Perform steps s_0, s_1,... as follows: at step s_i, if L_{i-1} is lit, then switch L_i from on to off or vice versa, otherwise do nothing. Show that:

(a) There is a positive integer $M(n)$ such that after $M(n)$ steps all the lamps are on again;

(b) If $n = 2^k$, then we can take $M(n) = n^2 - 1$.

(c) If $n = 2^k + 1$ then we can take $M(n) = n^2 - n + 1$.

1994

1994/1. Let m and n be positive integers. Let a_1, a_2, ..., a_m be distinct elements of $\{1, 2, \ldots, n\}$ such that whenever $a_i + a_j \leq n$ for some i, j (possibly the same) we have $a_i + a_j = a_k$ for some k. Prove that

$$\frac{a_1 + a_2 + \ldots + a_m}{m} \geq \frac{n+1}{2}.$$

1994/2. ABC is an isosceles triangle with $AB = AC$. M is the midpoint of BC and O is the point on the line AM such that OB is perpendicular to AB. Q is an arbitrary point on BC different from B and C. E lies on the line AB and F lies on the line AC such that E, Q and F are distinct and collinear. Prove that OQ is perpendicular to EF if and only if $QE = QF$.

1994/3. For any positive integer k, let $f(k)$ be the number of elements in the set $A_k = \{k+1, k+2, \ldots, 2k\}$ which have exactly three 1s when written in base 2. Prove that for each positive integer m, there is at least one k with $f(k) = m$ and determine all m for which there is exactly one k.

1994/4. Determine all ordered pairs (m, n) of positive integers for which

(1)
$$\frac{n^3 + 1}{mn - 1}$$

is an integer.

1994/5. Let S be the set of all real numbers greater than (-1). Find all functions f from S to S such that $f(x + f(y) + xf(y)) = y + f(x) + yf(x)$ for all x and y, $\dfrac{f(x)}{x}$ is strictly increasing on each of the intervals $-1 < x < 0$ and $0 < x$.

1994/6. Show that there exists a set A of positive integers with the following property: for any infinite set of primes, there exist two positive integers m in A and n not in A, each of which is a product of k distinct elements of S for some $k \geq 2$.

1995

1995/1. Let A, B, C, D be four distinct points on a line, in that order. The circles with diameter AC and BD intersect at X and Y. Let P be a point on the line XY other than Z. The line CP intersects the circle with diameter AC at C and M, and the line BP intersects the circle with diameter BD at B and N. Prove that the lines AM, DN, XY are concurrent.

1995/2. Let a, b, c be positive real numbers with $abc = 1$. Prove that
$$\frac{1}{a^3(b+c)} + \frac{1}{b^3(c+a)} + \frac{1}{c^3(a+b)} \geq \frac{3}{2}.$$

1995/3. Determine all integers $n > 3$ for which there exist n points A_1, A_2, $\ldots$, A_n in the plane, no three collinear and real numbers r_1, r_2, $\ldots$, r_n such that for any distinct i, j, k, the area of the triangle $A_i A_j A_k$ is $r_i + r_j + r_k$.

1995/4. Find the maximum value of x_0 for which there exists a sequence $x_0, x_1, \ldots, x_{1995}$ of positive reals with $x_0 = x_{1995}$ such that for $i = 1, 2, \ldots, 1995$

(i)
$$x_0 = x_{1995};$$

(ii)
$$x_{i-1} + \frac{2}{x_{i-1}} = 2x_i + \frac{1}{x_i}$$

1995/5. Let $ABCDEF$ be a convex hexagon with $AB = BC = CD$ and $DE = EF = FA$ such that $\angle BCD = \angle EFA = 60°$. Suppose that G and H are points in the interior of the hexagon such that $\angle AGB = \angle DHE = 120°$. Prove that

(1)
$$AG + GB + GH + DH + HE \geq CF.$$

1995/6. Let p be an odd prime number. How many p-element subsets A of $\{1, 2, \ldots, 2p\}$ are there, the sum of whose elements is divisible by p?

1996

1996/1. We are given a positive integer r and a rectangular board divided into 20×12 unit squares. The following moves are permitted on the board: one can move from one square to another only if the distance between the centres of the two squares is $\sqrt{r}$. The task is to find a sequence of moves leading between two adjacent corners of the board which lie on the long side.

(a) Show that the task cannot be done if r is divisible by 2 or 3.

(b) Prove that the task is possible for $r = 73$.

(c) Can the task be done for $r = 97$?

1996/2. Let P be a point inside the triangle ABC such that

(1)
$$\angle APB - \angle ACB = \angle APC - \angle ABC.$$

Let D, E be the incentres of triangles APB, APC respectively. Show that AP, BD and CE meet at a point.

1996/3. Let S be the set of non-negative integers. Find all functions f from S to itself such that

(1)
$$f(m + f(n)) = f(f(m)) + f(n)$$

for all m, n.

1996/4. The positive integers a and b are such that $15a + 16b$ and $16a - 15b$ are both squares of positive integers. What is the least possible value that can be taken by the smaller of these two squares?

1996/5. Let $ABCDEF$ be a convex hexagon of perimeter p such that AB is parallel to DE, BC is parallel to EF and CD is parallel to FA. Let R_A, R_C, and R_E denote the circumradii of triangles FAB, BCD, DEF respectively.

Prove that

(1)
$$R_A + R_C + R_E \geq \frac{p}{2}.$$

1996/6. Let p, q, n be positive integers with $p + q < n$. Let x_0, x_1, $\ldots$, x_n be integers such that $x_0 = x_n = 0$, and for each $1 \leq i \leq n$, $x_i - x_{i-1} = p$ or $-q$. Show that there exist indices $i < j$ with $(i, j) \neq (0, n)$ such that $x_i = x_j$.

1997

1997/1. In the plane the points with integer coordinates are the vertices of unit squares. The squares are coloured alternately black and white as on a chessboard. For any pair of positive integers m and n, consider a right-angled triangle whose vertices have integer coordinates and whose legs, of lengths m and n, lie along the edges of the squares. Let S_1 be the total area of the black part of the triangle, and S_2 be the total area of the white part. Let

$$f(m, n) = |S_1 - S_2|.$$

(a) Calculate $f(m, n)$ for all positive integers which are either both even or both odd.

(b) Prove that $f(m, n) \leq \frac{1}{2} \max(m, n)$ for all m, n.

(c) Show that there is no constant C such that $f(m, n) < C$ for all (m, n).

1997/2. The angle at A is the smallest angle in the triangle ABC. The points B and C divide the circumcircle of the triangle into two arcs. Let U be an interior point of the arc between B and C which does not contain A. The perpendicular bisectors of AB and AC meet the line AU at V and W, respectively. The lines BV and CW meet at T. Show that

$$AU = TB + TC.$$

1997/3. Let x_1, x_2, $\ldots$, x_n be real numbers satisfying

$$|x_1 + x_2 + \ldots + x_n| = 1$$

and

$$|x_i| \leq \frac{n+1}{2} \qquad (i = 1, 2, \ldots, n).$$

Show that there exists a permutation y_i of x_i such that

$$|y_1 + 2y_2 + 3y_3 + \ldots + ny_n| \leq \frac{n+1}{2}.$$

1997/4. An $n \times n$ matrix whose entries come from the set $S = \{1, 2, \dots$ $\dots, 2n-1\}$ is called silver matrix if, for each $i = 1, 2, \dots, n$, the ith row and the ith column together contain all elements of S. Show that:

(a) there is no silver matrix for $n = 1997$;

(b) silver matrices exist for infinitely many values of n.

1997/5. Find all pairs (a, b) of positive integers that satisfy

(1)
$$a^{b^2} = b^a.$$

1997/6. For each positive integer n, let $f(n)$ denote the number of ways of representing n as a sum of powers of 2 with non-negative integer exponents. Representations which differ only in the ordering of their summands are considered to be the same. For example, $f(4) = 4$, because 4 can be represented as

$$4; \quad 2+2; \quad 2+1+1; \quad 1+1+1+1.$$

Prove that for any integer $n \geq 3$

$$2^{\frac{n^2}{4}} < f(2^n) < 2^{\frac{n^2}{2}}.$$

1998

1998/1. In the convex quadrilateral $ABCD$, the diagonals AC and BD are perpendicular and the opposite sides AB and CD are not parallel. The point P, where the perpendicular bisectors of AB and CD meet, is inside $ABCD$. Prove that $ABCD$ is cyclic if and only if the triangles ABP and CDP have equal areas.

1998/2. In a competition there are a contestants and b judges, where $b \geq 3$ is an odd integer. Each judge rates each contestant as either "pass" or "fail". Suppose k is a number such that for any two judges their ratings coincide for at most k contestants. Prove

$$\frac{k}{a} \geq \frac{b-1}{2b}.$$

1998/3. For any positive integer n, let $d(n)$ denote the number of positive divisors of n (including 1 and n). Determine all positive integers k such that

$$\frac{d(n^2)}{d(n)} = k$$

for some n.

1998/4. Determine all pairs (a, b) of positive integers such that $(ab^2 + b + 7)$ divides $(a^2 b + a + b)$.

1998/5. Let I be the incentre of the triangle ABC. Let the incircle of ABC touch the sides BC, CA, AB at K, L, M, respectively. The line through B parallel to MK meets the lines LM and LK at R and S, respectively. Prove that the angle RIS is acute.

1998/6. Consider all functions f from the set of all positive integers into itself satisfying

(1)
$$f\left(t^2 f(s)\right) = s\,(f(t))^2$$

for all s and t. Determine the least possible value of $f(1998)$.

1999

1999/1. Find all finite sets S of at least three points in the plane such that for all distinct points A, B in S, the perpendicular bisector of AB is an axis of symmetry for S.

1999/2. Let $n \geq 2$ be a fixed integer. Find the smallest constant C such that for all non-negative reals $x_1, \cdots, x_n$:

(1)
$$\sum_{1 \leq i < j \leq n} x_i x_j (x_i^2 + x_j^2) \leq C \left(\sum_{1 \leq i \leq n} x_i \right)^4.$$

Determine when equality occurs.

1999/3. Given an $n \times n$ square board with n even. Two distinct squares of the board are said to be adjacent if they share a common side, but a square is not adjacent to itself. Find the minimum number of squares that can be marked so that every square (marked or not) is adjacent to at least one marked square.

1999/4. Find all pairs (n, p) of positive integers, such that p is a prime, $n \leq 2p$ and $(p-1)^n + 1$ is divisible by n^{p-1}.

1999/5. The circles C_1 and C_2 lie inside the circle C, and are tangent to it at M and N, respectively. C_1 passes through the centre of C_2. The common chord of C_1 and C_2, when extended, meets C at A and B. The lines MA and MB meet C_1 again at E and F. Prove that the line EF is tangent to C_2.

1999/6. Determine all functions $f : \mathbf{R} \to \mathbf{R}$ such that

(1)
$$f(x - f(y)) = f(f(y)) + xf(y) + f(x) - 1$$

for all x, y in $\mathbf{R}$. [$\mathbf{R}$ is the set of reals.]

2000

2000/1. Two circles Γ_1 and Γ_2 intersect at M and N. Let l be the common tangent to Γ_1 and Γ_2 so that M is closer to l than N is. Let l touch Γ_1 at A and Γ_2 at B. Let the line through M parallel to l meet the circle Γ_1 again at C and the circle Γ_2 again at D. Lines CA and DB meet at E; lines AN and CD meet at P; lines BN and CD meet at Q.

Show that $EP = EQ$.

2000/2. Let a, b, c be positive real numbers such that $abc = 1$. Prove that

$$(1) \qquad \left(a - 1 + \frac{1}{b}\right)\left(b - 1 + \frac{1}{c}\right)\left(c - 1 + \frac{1}{a}\right) \le 1.$$

2000/3. Let $n \ge 2$ be a positive integer. Initially, there are n fleas on a horizontal line, not all at the same point.

For a positive real number λ, define a *move* as follows:

choose any two fleas, at points A and B, with A to the left of B;

let the flea at A jump to the point C on the line to the right of B with $BC/AB = \lambda$.

Determine all the values of λ such that, for any point M on the line and any initial positions of the fleas, there is a finite sequence of moves that will take all the fleas to positions to the right of M.

2000/4. A magician has one hundred cards numbered 1 to 100. He puts them into three boxes, a red one, a white one and a blue one, so that each box contains at least one card.

A member of the audience selects two of the three boxes, choses one card from each and announces the sum of the numbers on the chosen cards. Given this sum, the magician identifies the box from which no card has been chosen.

How many ways are there to put all the cards into the boxes so that this trick always works? (Two ways are considered different if at least one card is put into a different box.)

2000/5. Determine whether or not there exists a positive integer n such that n is divisible by exactly 2000 prime numbers, and $2^n + 1$ is divisible by n.

2000/6. Let AH_1, BH_2, CH_3 be the altitudes of an acute-angled triangle ABC. The incircle of the triangle ABC touches the sides BC, CA, AB at T_1, T_2, T_3, respectively. Let the lines l_1, l_2, l_3 be the reflections of the lines H_2H_3, H_3H_1, H_1H_2 in the lines T_2T_3, T_3T_1, T_1T_2, respectively.

Prove that l_1, l_2, l_3 determine a triangle whose vertices lie on the incircle of the triangle ABC.

2001

2001/1. Let ABC be an acute-angled triangle with circumcentre O. Let P on BC be the foot of the altitude from A. Suppose that $BCA\angle \geq ABC\angle + 30°$. Prove that $CAB\angle + COP\angle < 90°$.

2001/2. Prove that

(1)
$$\frac{a}{\sqrt{a^2+8bc}} + \frac{b}{\sqrt{b^2+8ca}} + \frac{c}{\sqrt{c^2+8ab}} \geq 1$$

for all positive real numbers a, b, c.

2001/3. Twenty-one girls and twenty-one boys took part in a mathematical contest.

(1) Each contestant solved at most six problems.

(2) For each girl and each boy, at least one problem was solved by both of them.

Prove that there was a problem that was solved by at least three girls and at least three boys.

2001/4. Let n be an odd integer greater than 1, and let $k_1, k_2, \ldots, k_n$ be given integers. For each of the $n!$ permutations $a = (a_1, a_2, \ldots, a_n)$ of $1, 2, \ldots, n$ let

$$S(a) = \sum_{i=1}^{n} k_i a_i.$$

Prove that there are two permutations b and c, $b \neq c$, such that $n!$ is a divisor of $(S(b) - S(c))$.

2001/5. In a triangle ABC, let AP bisect $BAC\angle$, with P on BC, and let BQ bisect $ABC\angle$, with Q on CA. oldalon van. It is known that $BAC\angle = 60°$ and that $AB + BP = AQ + QB$.

What are the possible angles of triangle ABC?

2001/6. Let a, b, c, d be integers with $a > b > c > d > 0$. Suppose that

(1)
$$ac + bd = (b + d + a - c)(b + d - a + c).$$

Prove that $ab + cd$ is not a prime.

2002

2002/1. Let n be a positive integer. Let T be the set of points (x, y) in the plane where x and y are non-negative integers and $x + y < n$. Each point of T is coloured red or blue. If a point (x, y) is red then so are all points (x', y') of T with both $x' \leq x$ and $y' \leq y$. Define an X-set to be a set of n blue points having distinct x-coordinates and a Y-set to be a set of n blue points having distinct y-coordinates. Prove that the number of X-sets is equal to the number of Y-sets.

2002/2. Let BC be a diameter of the circle Γ with centre O. Let A be a point on Γ such that $0° < AOB\angle < 120°$. Let D be the midpoint of the arc $\overarc{AB}$ not containing C. The line through O parallel to DA meets the line AC at J. The perpendicular bisector of OA meets Γ at E and F. Prove that J is the incentre of the triangle CEF.

2002/3. Find all pairs of positive integers $m, n \geq 3$ such that there exist infinitely many positive integers a for which

$$\frac{a^m + a - 1}{a^n + a^2 - 1}$$

is an integer.

2002/4. Let n be an integer greater than 1. The positive divisors of n are $d_1, d_2, \ldots, d_k$ where

$$1 = d_1 < d_2 < \ldots < d_k = n.$$

Define $D = d_1 d_2 + d_2 d_3 + \ldots + d_{k-1} d_k$.

(a) Prove that $D < n^2$.

(b) Determine all n for which D is a divisor of n^2.

2002/5. Find all functions f from the set **R** of real numbers to itself such that

(1) $$(f(x) + f(z))(f(y) + f(t)) = f(xy - zt) + f(xt + yz)$$

for all x, y, z, t in **R**.

2002/6. Let $\Gamma_1, \Gamma_2, \ldots, \Gamma_n$ be circles of radius 1 in the plane, where $n \geq 3$. Denote their centres by $O_1, O_2, \ldots, O_n$, respectively. Suppose that no line meets more than two of the circles. Prove that

$$\sum_{1 \leq i < j \leq n} \frac{1}{O_i O_j} \leq \frac{(n-1)\pi}{4}.$$

2003

2003/1. Let A be a subset of the set $S = \{1, 2, \ldots, 1000000\}$ containing exactly 101 elements. Prove that there exist numbers t_1, t_2, $\ldots$, t_{100} in S such a way that the sets

$$A_j = \{x + t_j \mid x \in A\} \qquad j = 1, 2, \ldots, 100$$

are pairwise disjoint.

2003/2. Determine all pairs of positive integers (a, b) such that

$$\frac{a^2}{2ab^2 - b^3 + 1}$$

is a positive integer.

2003/3. A convex hexagon is given in which any two opposite sides have the following property: the distance between their midpoints is $\sqrt{3}/2$ times the sum of their lengths. Prove that all the angles of the hexagon are equal.

(A convex hexagon $ABCDEF$ has three pairs of opposite sides: AB and DE, BC and EF, CD and FA.)

2003/4. Let $ABCD$ a cyclic quadrilateral. Let P, Q and R be the feet of the perpendiculars from D to the lines BC, CA and AB respectively. Show that $PQ = QR$ if and only if the bisectors of $ABC\angle$ and $ADC\angle$ meet on AC.

2003/5. Let n be a positive integer and $x_1, x_2, \ldots, x_n$ be real numbers with $x_1 \leq x_2 \leq \ldots \leq x_n$.

(a) Prove that

$$\left(\sum_{i=1}^{n} \sum_{j=1}^{n} |x_i - x_j| \right)^2 \leq \frac{2(n^2 - 1)}{3} \sum_{i=1}^{n} \sum_{j=1}^{n} (x_i - x_j)^2.$$

(b) Show that equality holds if and only if $x_1, \ldots, x_n$ is an arithmetic sequence.

2003/6. Let p be a prime number. Prove that there exists a prime number q such that for any integer n, the number $n^p - p$ is not divisible by q.

2004

2004/1. Let ABC be an acute-angled triangle with $AB \neq AC$. The circle with diameter BC intersects the sides AB and AC at M and N respectively. Denote by O the midpoint of the side BC. The bisectors of the angles $BAC\angle$ and $MON\angle$ intersect at R. Prove that the circumcircles of the triangles BMR and CNR have a common point lying on the side BC.

2004/2. Find all polynomials $P(x)$ with real coefficients such that for all reals a, b, c such that $ab+bc+ca=0$ we have the following relation

(1) $$P(a-b)+P(b-c)+P(c-a)=2P(a+b+c).$$

2004/3. Define a *hook* to be a figure made up of six unit squares as shown below on the diagram, or any of the figures obtained by applying rotations and reflections to this figure.

Determine all $m \times n$ rectangles that can be covered with hooks such that

- the covering is without gaps and without overlaps,
- no part of a hook covers area outside the rectangle.

2004/4. Let $n \geq 3$ be an integer. Let t_1, t_2, ..., t_n be positive real numbers such that

$$n^2 + 1 > (t_1 + t_2 + \ldots + t_n)\left(\frac{1}{t_1} + \frac{1}{t_2} + \ldots + \frac{1}{t_n}\right).$$

Show that t_i, t_j, t_k are side lengths of a triangle for all i, j, k with $1 \leq i < < j < k \leq n$.

2004/5. In a convex quadrilateral $ABCD$ the diagonal BD bisects neither the angle $ABC\angle$, nor the angle $CDA\angle$. A point P lies inside $ABCD$ and satisfies

$$PBC\angle = DBA\angle \quad \text{and} \quad PDC\angle = BDA\angle.$$

Prove that $ABCD$ is a cyclic quadrilateral if and only if $AP = CP$.

2004/6. We call a positive integer *alternating* if every two consecutive digits in its decimal representation are of different parity.

Find all positive integers n such that n has a multiple which is alternating.

Solutions

1991.

1991/1. *Given a triangle ABC, let I be the incentre. The internal bisectors of angles A, B and C meet the opposite sides at A', B', C' respectively. Prove that*

(1)
$$\frac{1}{4} < \frac{AI \cdot BI \cdot CI}{AA' \cdot BB' \cdot CC'} \leq \frac{8}{27}.$$

First solution. Denote the area of the triangle by t, and that of the triangles ABI, BCI, CAI by t_C, t_A, t_B respectively; clearly $t = t_A + t_B + t_C$. Having BC as a common side the respective areas of the triangles BCI and BCA are in the ratio of the corresponding altitudes, $IA' : AA'$ (*Figure 1991/1.1*).

$$\frac{t_A}{t} = \frac{IA'}{AA'} = \frac{AA' - AI}{AA'} = 1 - \frac{AI}{AA'}$$

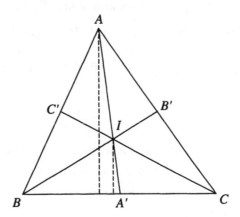

Figure 91/1.1

Thus

$$\frac{AI}{AA'} = 1 - \frac{t_A}{t},$$

and similarly

$$\frac{BI}{BB'} = 1 - \frac{t_B}{t}, \qquad \frac{CI}{CC'} = 1 - \frac{t_C}{t}.$$

By the A.M.–G.M. inequality

$$\frac{AI \cdot BI \cdot CI}{AA' \cdot BB' \cdot CC'} \leq \frac{1}{27}\left(\frac{AI}{AA'} + \frac{BI}{BB'} + \frac{CI}{CC'}\right)^3 = \frac{1}{27}\left(3 - \frac{t_A + t_B + t_C}{t}\right)^3 = \frac{8}{27}.$$

The second inequality is hence proved and observe that apart from being interior to the triangle ABC nothing else was used about the point I.

Turning to the proof of the first inequality note that $\dfrac{AI}{AA'} = 1 - \dfrac{t_A}{t} = \dfrac{t_B + t_C}{t}$;
if r is the inradius then $2t = r(a+b+c)$, $2t_A = ra$, $2t_B = rb$, $2t_C = rc$ and hence
$\dfrac{AI}{AA'} = \dfrac{b+c}{a+b+c}$; similarly

$$\frac{BI}{BB'} = \frac{c+a}{a+b+c}, \qquad \frac{CI}{CC'} = \frac{a+b}{a+b+c}$$

and thus the claim becomes

(2) $\qquad\qquad 4(a+b)(b+c)(c+a) > (a+b+c)^3.$

Performing the calculations and rearranging we get

$$-a^3 - b^3 - c^3 + 2a^2 b + ab^2 + a^2 c + ac^2 + b^c + bc^2 + 2abc > 0.$$

which is the same as

(3) $\qquad\qquad (a+b-c)(a+c-b)(b+c-a) + 4abc > 0$

and (3) obviously holds if a, b and c are the sides of a triangle.

Second solution. The bisector of the interior angle divides the opposite side
into the ratio of the neighbouring sides and hence $BA' = \dfrac{ac}{b+c}$, for example; applying this theorem for the bisector BI of $\triangle ABA'$ yields

$$AI = \frac{c \cdot AA'}{c + \frac{ac}{b+c}} = \frac{(b+c) \cdot AA'}{a+b+c}, \quad \text{and thus} \quad \frac{AI}{AA'} = \frac{b+c}{a+b+c}.$$

Similarly

$$\frac{BI}{BB'} = \frac{a+c}{a+b+c}, \qquad \frac{CI}{CC'} = \frac{a+b}{a+b+c}.$$

The double inequality (1) hence can be written as follows:

(4) $\qquad\qquad \dfrac{1}{4} < \dfrac{(a+b)(b+c)(c+a)}{(a+b+c)^3} \leq \dfrac{8}{27}.$

By the A.M.–G.M. inequality:

$$\frac{1}{(a+b+c)^3} \cdot (a+b)(b+c)(c+a) \leq \frac{1}{(a+b+c)^3} \left(\frac{a+b+b+c+c+a}{3} \right)^3 = \frac{8}{27},$$

which is the second inequality.

Using the identity

$$3(a+b)(b+c)(c+a) = (a+b+c)^3 - a^3 - b^3 - c^3$$

the first inequality becomes

(5) $\qquad\qquad \dfrac{1}{4} < \dfrac{(a+b+c)^3 - a^3 - b^3 - c^3}{3(a+b+c)^3}.$

To prove this we use the following *Lemma.* If x, y, x_1, y_1 are positive numbers such that $x + y = x_1 + y_1$ and $x - y < x_1 - y_1$ ($x \geq y$, $x_1 \geq y_1$) then
$x^3 + y^3 < x_1^3 + y_1^3$.

When applying the lemma $a \geq b \geq c$ can clearly be assumed. The cast will be

$$x = a, \quad y = b, \quad x_1 = \frac{a+b+c}{2}, \quad y_1 = \frac{a+b-c}{2}.$$

The conditions of the lemma now hold: $x \geq y$, $x_1 \geq y_1$, $x + y = x_1 + y_1$ and $x - y = a - b \leq c = x_1 - y_1$ so

$$a^3 + b^3 < \left(\frac{a+b+c}{2}\right)^3 + \left(\frac{a+b-c}{2}\right)^3.$$

Using this and also the easy to verify inequality $\left(\frac{a+b-c}{2}\right)^3 + c^2 < \left(\frac{a+b+c}{2}\right)^3$

(5) becomes

$$\frac{(a+b+c)^3 - a^3 - b^3 - c^3}{3(a+b+c)^3} > \frac{(a+b+c)^3 - \left(\frac{a+b+c}{2}\right)^3 - \left(\frac{a+b-c}{2}\right)^3 - c^3}{3(a+b+c)} >$$

$$> \frac{(a+b+c)^3 - \left(\frac{a+b+c}{2}\right)^3 - \left(\frac{a+b+c}{2}\right)^3}{3(a+b+c)^3} = \frac{1}{4},$$

indeed.

Remarks. 1. Both solutions show that equality holds in the second inequality if our triangle is equilateral.

In the first inequality the lower bound $\frac{1}{4}$ cannot be improved. If $b = c$ are fixed and a tends to zero then the limit of the ratio in (4) is $\frac{1}{4}$ and thus the ratio can be arbitrarily close to this number.

2. Let's prove now the lemma of the second solution.

From the conditions $x^2 + 2xy + y^2 = x_1^2 + 2x_1y_1 + y_1^2$ and $x^2 - 2xy + y^2 < x_1^2 - 2x_1y_1 + y_1^2$. Their difference is $-4xy < -4x_1y_1$ that is $xy > x_1y_1$. Raising the equality in the condition to the third power

$$x^3 + y^3 + 3xy(x+y) = x_1^3 + y_1^3 + 3x_1y_1(x_1+y_1),$$

$$0 = x_1^3 + y_1^3 - (x^3 + y^3) + 3(x+y)(x_1y_1 - xy),$$

$$x_1^3 + y_1^3 - (x^3 + y^3) > 0,$$

and this was to be proved.

1991/2. *Let $n > 6$ be an integer and let a_1, a_2, ..., a_k be all the positive integers less than n and relatively prime to n. If*

$$a_2 - a_1 = a_3 - a_2 = \ldots = a_k - a_{k-1} > 0,$$

prove that n must be either a prime number or a power of 2.

Solution. Rephrasing the problem: if the numbers less than n and prime to it form an arithmetic progression then n itself is either a prime or it is a power of 2.

The common difference of the A.P. is a positive integer, denote it by d. Since 1 and $n-1$ are both prime to n, $a_1 = 1$ and $a_k = n-1$.

Let n be odd first. Then $n-2$ is also prime to n. Indeed, their g.c.d. divides 2, their difference, but n as an odd number is not divisible by 2. Hence $d=1$, the A.P. consists of the integers from 1 to $n-1$, these numbers are all prime to n which, with no proper divisors, is a prime number, indeed.

Let n be now even; then it can be written as $n = 2^r \cdot s$ where $r \geq 1$ integer and $s \geq 1$ is an odd integer.

If $s = 1$ then n is a power of 2, the numbers prime to n are the odd integers and they form an A.P., indeed. It is left to show that $s > 1$ is impossible.

α) If $s = 3$ then $n > 6$ and thus the sequence of the numbers prime to n begins with as 1, 5, 7, $\ldots$, which is certainly not an arithmetic progression.

β) For $s \geq 5$ we show that $s-2$ and $s-4$ are both terms of the $A.P.$ of the problem. For the proof consider an arbitrary common divisor, $c \geq 1$, of $n = 2^r \cdot s$ and $s-2$. Since $2^r \cdot s = 2^r(s-2) + 2^{r+1}$, it also divides 2^{r+1}. On the other hand, as a divisor of the odd $s-2$, c is an odd divisor of a power of 2; thus $c = 1$. Hence n and $s-2$ are coprime, $s-2$ belongs to the $A.P.$, indeed. $2^r(s-4) + 2^{r+2} = 2^r \cdot s = n$ now implies that the $g.c.d.$ of $s-4$ and n is 1 and thus $s-4$ is in the $A.P.$ as well. With $s-4$ and $s-2$ the number s is also in the arithmetic progression of the condition; this, however, is a contradiction since s was a proper divisor of n; the proof is complete.

1991/3. *Let* $S = \{1, 2, 3, \ldots, 280\}$. *Find the smallest integer* n *such that each* n-*element subset of* S *contains five numbers which are pairwise relatively prime.*

Solution. First we show a 216-element subset of S which does not contain five pairwise relatively prime numbers. Pick the multiples of 2, 3, 5 and 7 from S and denote the subset thus obtained by N. Denote also by N_i the multiples of i in S and its cardinality by the usual $|N_i|$.

Clearly

$$|N_2| = 140, \quad |N_3| = 93, \quad |N_5| = 56, \quad |N_7| = 40.$$

$|N_2 \cap N_3|$, for example, is the number of the common multiples of 2 and 3, that is the multiples of 6; for these sets

$$|N_2 \cap N_3| = 46, \quad |N_2 \cap N_5| = 28, \quad |N_2 \cap N_7| = 20,$$
$$|N_3 \cap N_5| = 18, \quad |N_5 \cap N_7| = 8, \quad |N_3 \cap N_7| = 13.$$

We also need

$$|N_2 \cap N_3 \cap N_5| = 9, \ |N_2 \cap N_3 \cap N_7| = 6, \ |N_2 \cap N_5 \cap N_7| = 4, \ |N_3 \cap N_5 \cap N_7| = 2$$
$$|N_2 \cap N_3 \cap N_5 \cap N_7| = 1.$$

By the principle of inclusion and exclusion

$$|N| = (140 + 93 + 56 + 40) - (46 + 28 + 20 + 18 + 8 + 13) + (9 + 6 + 4 + 2) - 1 = 216.$$

Choosing now any five elements of N at least one of the four subsets N_2, N_3, N_5 and N_7 will be represented twice and these two numbers are not coprime, indeed.

Next we prove that no matter how do we select 217 numbers from S there are 5 pairwise relatively prime ones among the selected numbers; the smallest value of n we are looking for is hence 217.

There are 59 primes among the first 280 positive integers (cf. the remark at the end of this solution). Adjoining also 1 to these 59 numbers we get a 60-element set, denote it by P. Now there are 220 composite numbers in S; they form the set T. We now list 8 disjoint subsets of T, each comprising five pairwise relatively prime elements:

$$M_1 = \{2 \cdot 23, \ 3 \cdot 19, \ 5 \cdot 17, \ 7 \cdot 13, \ 11^2\},$$
$$M_2 = \{2 \cdot 29, \ 3 \cdot 23, \ 5 \cdot 19, \ 7 \cdot 17, \ 11 \cdot 13\},$$
$$M_3 = \{2 \cdot 31, \ 3 \cdot 29, \ 5 \cdot 23, \ 7 \cdot 19, \ 11 \cdot 17\},$$
$$M_4 = \{2 \cdot 37, \ 3 \cdot 31, \ 5 \cdot 29, \ 7 \cdot 23, \ 11 \cdot 19\},$$
$$M_5 = \{2 \cdot 41, \ 3 \cdot 37, \ 5 \cdot 31, \ 7 \cdot 29, \ 11 \cdot 23\},$$
$$M_6 = \{2 \cdot 43, \ 3 \cdot 41, \ 5 \cdot 37, \ 7 \cdot 31, \ 13 \cdot 17\},$$
$$M_7 = \{2 \cdot 47, \ 3 \cdot 43, \ 5 \cdot 41, \ 7 \cdot 37, \ 13 \cdot 19\},$$
$$M_8 = \{2^2, \ 3^2, \ 5^2, \ 7^2, \ 13^2\}.$$

Consider now an arbitrary 217-element subset H of S. If H and P have at least 5 common elements then we are done, these 5 numbers are pairwise coprime.

We may hence assume that there are at most 4 elements of P in H and thus its further 213 elements are all composite. This means that there are at most 7 missing from the altogether 220 composite numbers in S and thus one of the sets M_i is contained in H. This M_i comprises the 5 numbers required.

Remark. There are no calculators or tables available at the *IMO*. With such aids around the number of primes below 280 can be found by either the *sieve* of *Eratosthenes* or with bare hands by removing the primes 2, 3, 5, 7 from the set N leaving 212 composite numbers; apart from these there are only 8 composite elements in S: 11^2, 13^2, $11 \cdot 13$, $11 \cdot 17$, $11 \cdot 19$, $11 \cdot 23$, $13 \cdot 17$ and $13 \cdot 19$.

1991/4. *Suppose G is a connected graph with k edges. Prove that it is possible to label the edges 1, 2, 3, ..., k in such a way that at each vertex which belongs to two or more edges, the greatest common divisor of the integers labelling those edges is 1.*

[A graph is a set of points, called vertices, together with a set of edges joining certain pairs of distinct vertices. Each pair of edges belongs to at most one edge. The graph is connected if for each pair of distinct vertices x, y there is some sequence of vertices $x = v_0, v_1, \ldots, v_m = y$, such that each pair v_i, v_{i+1} $(0 \le i < m)$ is joined by an edge.]

First solution. In the solution we shall try to do the labelling in such a way that each vertex belongs to some pair of edges labelled by consecutive and hence coprime numbers.

The degree of a vertex is the number of edges leaving from that vertex. Denote the vertices by $P_1, P_2, \ldots, P_n$. Since the graph is connected there is a path between any two vertices. Start hence from P_1 to P_2, go on to P_3 and so on up to P_n, finally, from here go back to P_1. Label the edges crossed along the roundtrip as follows:

a) the first edge of the trip starting from P_1 is labelled by 1;

b) travelling along an already labelled edge its label remains unaltered; numbering is resumed only when arriving to an edge not labelled so far;

c) when arriving for the first time to a vertex whose degree is not 1 continue the trip along any edge not yet labelled; this edge gets labelled now and with the number one greater than the edge of arrival;

d) having entered the numbers 1, 2, ..., s by the end of the trip if there are yet unlabelled edges then these are numbered arbitrarily by the numbers $s + 1$, $s + 2$, ..., k.

This labelling has the desired property: indeed, apart from the first degree vertices and maybe P_1 every edge belongs to at least one pair of edges labelled by consecutive numbers written upon the first arrival to this vertex; the edge labelled by 1 belongs to P_1 which settles this vertex as well; finally, the first degree vertices are clearly irrelevant.

Second solution. It is well known that in a graph the number of odd degree vertices is even. Arrange the odd degree vertices into pairs and connect, in each pair, the vertices by a red edge. Colour also the edges in G by black. Each vertex of the graph G' prepared this way is of even degree. There might be multiple edges (a black and a new red one) in G' but this does not matter. Note that there is at most one red edge belonging to any vertex.

Graph G' is well known to be *Eulerian*. Starting from any vertex the edges of such graphs can be traversed along a closed path without retracing any edge, along a so called *Euler-circuit*.

Traverse now the edges of G' along an Euler-circuit. Label the first black edge crossed by 1 and each subsequent black edge of the circuit is labelled by the next integer. Red edges are not numbered at all. Since the circuit contains each of them, every edge in G is get labelled. If a vertex belongs to black edges only then its degree in G is even and thus there are consecutive ones among the labels of these edges. If there is a red edge here and a single black one then its label has certainly no effect. Finally, if there are more black edges then, by construction, there are at least three of them, the circuit is passing through this vertex at least twice and on one of these occasions both the arriving and the leaving edges are black. These edges hence are labelled by consecutive numbers as required.

1991/5. *Let ABC be a triangle and X an interior point of ABC. Show that at least one of the angles XAB, XBC, PCA is less than or equal to $30°$.*

First solution. We shall use the notations of *Figure 1991/5.1.*

If some angle of the triangle is at least $150°$ then there is another one strictly less than $30°$ and thus the claim obviously holds; assume now, indirectly, that $30° < A_1$, B_1, $C_1 < 150°$. Hence, if the feet of the perpendiculars from P to AB, BC, CA are C', A', B', respectively, then

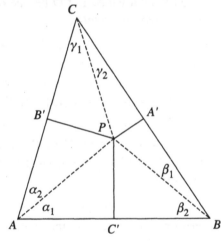

Figure 91/5.1

$$\sin A_1 = \frac{PC'}{PA} > \sin 30° = \frac{1}{2},$$

that is

$$2PC' > PA$$

and similarly

$$2PA' > PB, \qquad 2PB' > PC.$$

Thus

(1)
$$PA' + PB' + PC' > \frac{PA + PB + PC}{2},$$

contradicting the assertion of the *Erdős–Mordell inequality* ([33]) where, instead of '>', (1) holds with '≤'. This reference completes the proof.

Second solution. Assume again that A_1, B_1, C_1 are greater than $30°$ which, like in the first solution, implies that $30° < A_1$, B_1, $C_1 < 150°$. Thus

(2)
$$\sin A_1 \sin B_1 \sin C_1 > \frac{1}{8}.$$

By the assumption $A_1 + B_1 + C_1 > 90°$ and hence $A_2 + B_2 + C_2 < 90°$. Applying now the A.M.–G.M. inequality and also *Jensen's inequality*

$$\sin A_2 \sin B_2 \sin C_2 \le \left(\frac{\sin A_2 + \sin B_2 + \sin C_2}{3}\right)^3 \le$$

$$\leq \sin^3 \frac{A_2 + B_2 + C_2}{3} < \sin^3 30° = \frac{1}{8},$$

that is

$$\frac{1}{\sin A_2 \sin B_2 \sin C_2} > 8,$$

which, when combined with (2), yields

(3) $$\frac{\sin A_1 \sin B_1 \sin C_1}{\sin A_2 \sin B_2 \sin C_2} > 1.$$

Applying now the sine rule in the triangles APB, BPC, CPA respectively:

$$\frac{\sin A_1}{\sin B_2} \cdot \frac{\sin B_1}{\sin C_2} \cdot \frac{\sin C_1}{\sin A_2} = \frac{PB}{PA} \cdot \frac{PC}{PB} \cdot \frac{PA}{PC} = 1,$$

contradicting to (3), the consequence of our assumption. There must be at least one among A_1, B_1, C_1, less than $30°$, indeed.

Third solution. Let Q be one of the triangle's *Brocard's points* ([34]). Now with the *Brocard angle* ω

$$QBC\angle = QCA\angle = QAB\angle = \omega.$$

Since

$$\cot \omega = \cot A + \cot B + \cot C,$$

the *cotangent inequality* ([3]) implies

$$\cot \omega \geq \sqrt{3},$$

that is $\omega \leq 30°$. Point P of the problem is contained by one of the triangles QBC, QCA, QAB, say QBC; hence $\angle PBC \leq \angle QBC = \omega \leq 30°$, and this was to be proved.

Remarks. 1. If the respective distances from the vertices of a triangle of the interior point P are p, q and r and from the sides are x, y and z then

$$p + q + r \geq 2(x + y + z).$$

This is the *Erdős–Mordell inequality*. Equality holds if and only if the triangle is equilateral and P is its centre.

2. Points Q_1 and Q_2 are the *Brocard's points* of $\triangle ABC$ if

$$\angle BAQ_1 = \angle CBQ_1 = \angle ACQ_1, \quad \text{and} \quad \angle ABQ_2 = \angle BCQ_2 = \angle CAQ_2.$$

Apart from the circumcentre these are the only points with the property that their perpendicular projections on the sides form a triangle similar to the original one. This also implies that if the triangle is rotated about any one of it Brocard-points then the triangles formed by the intersections of the corresponding pairs of sides are similar.

3. The cotangent inequality states that in an arbitrary triangle

$$\cot A + \cot B + \cot C \geq \sqrt{3}.$$

Equality holds if and only if the triangle is equilateral.

1991/6. *Given any real number $a > 1$ construct a bounded infinite sequence $x_0, x_1, \ldots$ such that*

(1) $$|x_i - x_j| \cdot |i - j|^a \geq 1$$

for every pair of distinct i, j. [An infinite sequence $x_0, x_1, \ldots$ of real numbers is bounded if there is a constant C such that $|x_i| < C$ for all i.]

First solution. Note first that if (1) holds with $a = 1$ for every pair of distinct i, j then it clearly holds for $a > 1$. Being so it is enough to construct a sequence satisfying (1) if $a = 1$. In due course we shall make use of the well known result about the error when $\sqrt{2}$ is approximated by fractions of the form $\frac{p}{q}$ (p and q are positive integers):

(2) $$\left| \sqrt{2} - \frac{p}{q} \right| > \frac{1}{3q^2}.$$

Denote now the fractional part of $\sqrt{2}k$ by the usual $\{\sqrt{2}k\}$ and let

(3) $$x_k = 3\{\sqrt{2}k\} = 3\left(\sqrt{2}k - [\sqrt{2}k]\right), \qquad k = 0, 1, 2, \ldots$$

We show that this sequence satisfies (1) with $a = 1$. Let $i > j$ be non negative integers. Since $\{\sqrt{2}k\} < 1$,

$$x_k < 3,$$

the sequence is bounded. (1) also holds because if $q = i - j$ and $p = [i\sqrt{2}] - [j\sqrt{2}]$ then

$$|x_i - x_j||i - j| = 3\left|(i - j)\sqrt{2} - \left([i\sqrt{2}] - [j\sqrt{2}]\right)\right|(i - j) =$$

$$= 3|q\sqrt{2} - p|q = 3q^2 \left|\sqrt{2} - \frac{p}{q}\right| > 1,$$

and the latter is true by (2). The sequence (3) hence satisfies the requirements for every $a \geq 1$.

Second solution. The sequence will be defined in terms of the binary form of natural numbers.

Let the binary form of i be:

$$i = b_0 + b_1 \cdot 2 + b_2 \cdot 2^2 + \ldots + b_k \cdot 2^k,$$

where b_s; the binary digits, are 0 or 1. With the given value of $a \geq 1$ prepare the following real number h_i from the digits of i:

$$h_i = b_0 + b_1 \cdot 2^{-a} + b_2 \cdot 2^{-2a} + \ldots + b_k \cdot 2^{-ka}.$$

As the sum of an infinite *G.P.*:

(4) $$0 \leq h_i \leq 1 + 2^{-a} + 2^{-2a} + \ldots + a^{-ka} < \frac{1}{1 - 2^{-a}}.$$

Similarly, if j is different from i and $j = c_0 + c_1 \cdot 2 + c_2 \cdot 2^2 + \ldots + c_k \cdot 2^k$ then

$$h_j = c_0 + c_1 \cdot 2^{-a} + c_2 \cdot 2^{-2a} + \ldots + c_k \cdot 2^{-ka}.$$

Consider now the first binary position where i and j are different; if this is the pair (b_t, c_t) then $|i - j| \geq 2^t$. The following chain of estimates is now clearly valid:

$$\left| h_i - h_j \right| = \left| (b_0 - c_0) + (b_1 - c_1) \cdot 2^{-a} + \ldots + (b_k - c_k) 2^{-ka} \right| \geq$$

$$\geq |b_t - c_t| 2^{-ta} - |b_{t+1} - c_{t+1}| 2^{-(t+1)a} - \ldots - |b_k - c_k| 2^{-ka} \geq$$

$$\geq 2^{-ta} - 2^{-(t+1)a} - \ldots - 2^{-ka} > 2^{-ta} - \frac{2^{-(t+1)a}}{1 - 2^{-a}} =$$

$$= 2^{-ta} \left(1 - \frac{2^{-a}}{1 - 2^{-a}} \right) = \left(2^t \right)^{-a} \frac{2^a - 2}{2^a - 1} \geq \frac{2^a - 2}{2^a - 1} |i - j|^{-a}.$$

Rearranging we arrive to

(5) $$\left| h_i - h_j \right| |i - j|^a \geq \frac{2^a - 2}{2^a - 1}.$$

Define now the sequence as:

$$x_i = \frac{2^a - 1}{2^a - 2} h_i.$$

We show that this sequence satisfies the requirements. First of all, by (4)

$$x_i < \frac{2^a - 2}{2^a - 1} \cdot \frac{a}{1 - 2^{-a}} = a \frac{2^a - 2}{2^a - 2 + 2^{-a}} < a,$$

the sequence is hence bounded. As for (1) we clearly have

$$|x_i - x_j| |i - j|^a = \frac{2^a - 1}{2^a - 2} |h_i - h_j| |i - j|^a \geq \frac{2^a - 1}{2^a - 2} \cdot \frac{2^a - 2}{2^a - 1} = 1,$$

and the desired property is immediate.

Remark. The method of the first solution when improved yields much better estimations for the terms of the sequence.

The result about the error of the approximating fractions of $\sqrt{2}$ can be proved as follows. Once more, it states that if $\frac{p}{q}$ is a positive rational number then

$$\left| \sqrt{2} - \frac{p}{q} \right| > \frac{1}{3q^2}.$$

If $q = 1$ then this is obvious so $q \geq 2$ can be assumed. Suppose now the assertion is false and hence

$$\left| \sqrt{2} - \frac{p}{q} \right| \leq \frac{1}{3q^2}$$

holds for some $\frac{p}{q}$. This means that

$$-\frac{1}{3q^2} \leq \sqrt{2} - \frac{p}{q} \leq \frac{1}{3q^2}$$

which, when rearranged, becomes

$$-\frac{1}{3q}+q\sqrt{2}\leq p\leq\frac{1}{3q}+q\sqrt{2}$$

This implies

$$-\frac{2\sqrt{2}}{3}+\left(\frac{1}{3q}\right)^2\leq p^2-2q^2\leq\frac{2\sqrt{2}}{3}+\left(\frac{1}{3q}\right)^2,$$

and by the assumption $q\geq 2$ we arrive to

$$|p^2-2q^2|\leq\frac{2\sqrt{2}}{3}+\left(\frac{1}{3q}\right)^2\leq\frac{2\sqrt{2}}{3}+\frac{1}{36}=\frac{24\sqrt{2}+1}{36}<1.$$

Since $|p^2-2q^2|$ is a non negative integer, the only possibility is $p^2-2q^2=0$,

$$\frac{p}{q}=\sqrt{2}$$

contradicting the irrationality of $\sqrt{2}$. The proof is finished.

1992.

1992/1. *Find all integers* a, b, c *satisfying* $1<a<b<c$ *such that* $(a-1)\cdot(b-1)\cdot(c-1)$ *is a divisor of* $abc-1$.

Solution. With

(1)
$$Q=\frac{abc-1}{(a-1)(b-1)(c-1)}$$

we have to find those integral values of a, b, c for which Q is an integer. Clearly

(1') $Q=\dfrac{abc}{(a-1)(b-1)(c-1)}-\dfrac{1}{(a-1)(b-1)(c-1)}=$

$$=\left(1+\frac{1}{a-1}\right)\left(1+\frac{1}{b-1}\right)\left(1+\frac{1}{c-1}\right)-\frac{1}{(a-1)(b-1)(c-1)},$$

therefore

(2) $Q<\left(1+\dfrac{1}{a-1}\right)\left(1+\dfrac{1}{b-1}\right)\left(1-\dfrac{1}{c-1}\right).$

(1) implies that if any one of a, b, c is odd then so is the numerator; thus, for Q to be an integer, the denominator has to be odd as well, that is a, b and c are either all odd or all even. We also note that $Q>1$ is immediate from (1').

Next we show that a cannot exceed 3. Indeed, from what we know of these numbers, $a\geq 4$ implies $b\geq 6$ and $c\geq 8$ and thus the *r.h.s.* of (2) is at most

$$\left(1+\frac{1}{4-1}\right)\left(1+\frac{1}{6-1}\right)\left(1+\frac{1}{8-1}\right)=\frac{192}{105},$$

less than 2. But then Q as a whole number is strictly between 1 and 2, a contradiction. The possible values of a are hence 2 and 3.

α) If $a = 2$, then both b and c are even, $b \geq 4$ and $c \geq 6$ and (2) implies

$$Q < \frac{16}{5} < 4$$

so either $Q = 2$ or $Q = 3$. The former is now impossible since, by (1), Q is odd; therefore, if $a = 2$ then $Q = 3$ and (1) becomes

$$2bc - 1 = 3(b - 1)(c - 1),$$
$$(b - 3)(c - 3) = 5 = 1 \cdot 5,$$

and thus $b = 4$ and $c = 8$. The triplet $(2, 4, 8)$ is a solution, indeed.

β) If $a = 3$ then proceeding as above, $b \geq 5$, $c \geq 7$; (2) now yields

$$Q < \frac{35}{16} < 3$$

that is $Q = 2$. Plugging these values in (1):

$$2 = \frac{3cb - 1}{2(b - 1)(c - 1)},$$
$$(b - 4)(c - 4) = 11 = 1 \cdot 11,$$

and thus $b = 5$ and $c = 15$. The triplet $(3, 5, 15)$ is the other solution of the problem.

1992/2. *Find all functions f defined on the set of all real numbers with real values, such that*

(1) $$f\left(x^2 + f(y)\right) = y + (f(x))^2$$

for all x, y.

In what follows the notation $f^2(x)$ will be used for $(f(x))^2$, the square of $f(x)$.

First solution. Guessing and checking $f(x) = x$ turns out to be a solution; we show that this is the only one.

Assume first that for some real number y

(2) $$f(y) < y, \qquad \text{that is} \qquad y - f(y) > 0$$

and let x be such that $x^2 = y - f(y)$ that is $y = x^2 + f(y)$. By (1), this implies

$$f(y) = f\left(x^2 + f(y)\right) = y + f^2(x) \geq y,$$

contradicting (2); hence for every real number y

(3) $$f(y) \geq y.$$

Set now y_0-t to be smaller than $-f^2(0)$ and denote $f(y_0)$ by a. Combining (3) and (1) we get

$$a \leq f(a) = f\left(0^2 + f(y_0)\right) = y_0 + f^2(0) < 0 \quad \text{that is}$$
$$a \leq f(a) < 0, \quad \text{and thus}$$

(4) $$a^2 \geq f^2(a).$$

Let x be now arbitrary. Using (3), (1) and (4) we arrive to

(5) $x + a^2 \le a^2 + f(x) \le f\left(a^2 + f(x)\right) = x + f^2(a) \le x + a^2.$

Since the lowest and the highest terms are equal, there is equality everywhere in (5) and thus

$$f(x) = x,$$

for every real number x, indeed.

Second solution. Let a be an arbitrary real number. We show that f assumes the value a and the corresponding argument is $x^2 + f\left(a - f^2(x)\right)$ where x is arbitrary. For this set $y = a - f^2(x)$ in (1):

(6) $f\left(x^2 + f\left(a - f^2(x)\right)\right) = a - f^2(x) + f^2(x) = a,$

indeed. Next we show that f is one to one, that is $y_1 \ne y_2$ implies $f(y_1) \ne f(y_2)$.

Let x be once more an arbitrary real number. Applying (1) twice:

$$f\left(x^2 + f(y_1)\right) = y_1 + f^2(x),$$
$$f\left(x^2 + f(y_2)\right) = y_2 + f^2(x).$$

Thus $f(y_1) = f(y_2)$ implies $y_1 = y_2$, f is one to one, indeed, it is a bijection of the set of the real numbers.

Apply now (1) for each of the pairs (x, y) and $(-x, y)$:

$$f\left(x^2 + f(y)\right) = y + f^2(x),$$
$$f\left(x^2 + f(y)\right) = y + f^2(-x).$$

The difference of the respective sides can be factorized:

$$f^2(x) - f^2(-x) = 0,$$
$$(f(x) - f(-x))(f(x) + f(-x)) = 0.$$

Assume first that x is not zero. Then $f(x) = f(-x)$ is not possible because f is one to one. Hence

$$f(x) = -f(-x),$$

so, apart from $x = 0$, our function is odd. Now we show that $f(0) = 0$ also holds. If $a = x = 0$ then (6) becomes $f(f(0)) = 0$. Assume now that $f(0) \ne 0$. Then, as we have just shown, $f(-f(0)) = -f(f(0))$. But the latter is equal to zero, so if $f(0) \ne 0$ then $f(-f(0)) = 0$. So $f(f(0)) = f(-f(0))$ and thus $f(0) = -f(0)$ because f is one to one. The last equality means that $f(0) = 0$ that is $f(0) \ne 0$ is indeed impossible. Set now $x = 0$ in (1). Thus

$$(7) \qquad\qquad f(f(y)) = y.$$

Finally we prove that f is increasing that is if $x_1 \le x_2$ then $f(x_1) \le f(x_2)$. Let $x_2 = x_1 + x_0^2$ ($x_0 \ge 0$). Setting $x = x_0$, $y = f(x_1)$ in (1) and using (7)

$$f(x_2) = f(x_0^2 + x_1) = f(x_0 + f(f(x_1))) = f(x_1) + f^2(x_0) \ge f(x_1).$$

It is the monotonity that forces f to be equal to the identity. Indeed, if $x > f(x)$ then, as an increasing function, $f(x) \ge f(f(x))$ which, by (7), is equal to x. Thus $x > f(x)$ yields $x \le f(x)$ and, similarly, from $x < f(x)$ we can deduce $x \ge f(x)$. Therefore the only possibility is $f(x) = x$ and this is indeed a solution.

1992/3. *Consider 9 points in space, no 4 coplanar. Each pair of points is joined by a line segment which is coloured either blue or red or left uncoloured. Find the smallest value of n such that whenever exactly n edges are coloured, the set of coloured edges necessarily contains a triangle all of whose edges have the same colour.*

Solution. The problem when rephrased in usual graph-terminology is as follows: at least how many edges of a complete graph with nine vertices have to be coloured with red or blue if for any colouring there is a monochromatic triangle.

We prove that the number we are looking for is 33. The complete nine-point graph has $\binom{9}{2} = 36$ edges. Delete 3 of them randomly and colour the remaining 33 edges arbitrarily. Select 3 different ones among the endpoints of the deleted edges and delete them from the graph, together with the coloured edges adjacent to these erased vertices.

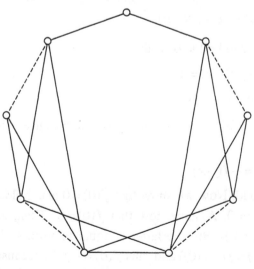

Figure 92/3.1

The remaining graph has 6 vertices and the edges connecting any two of them are all coloured; this one is a complete 6-graph. According to a *Ramsey-type* theorem, stated and proved in the solution of Problem 1964/4 one can always find a monochromatic triangle in such a graph. Therefore, any 9-point graph with 33 edges contains a monochromatic triangle, indeed.

The graph on *Figure 1992/3.1* shows a 9-point graph with 32 edges where there is no monochromatic triangle. Hence the minimal value of n having the given property is 33. There are the blue edges indicated on the figure, only, and you can check that they do not form a triangle, indeed. Among any three vertices, however, there are two connected by either a blue edge or an uncoloured one and thus there is no red triangle, either. (The four deleted edges are indicated on the figure by a dotted line).

Remark. According to a celebrated theorem of *Paul Turán* the problem can be extended to graphs of arbitrary size.

1992/4. *L is tangent to the circle C and M is a point on L. Find the locus of all points P such that there exist points Q and R on L equidistant from M with C the incircle of the triangle PQR.*

Solution. Consider a triangle PQR satisfying the conditions (*Figure 1992/4.1*). Circle C is touching l at E. G is the point diametric to E on the circle C and the tangent to C at G is e.

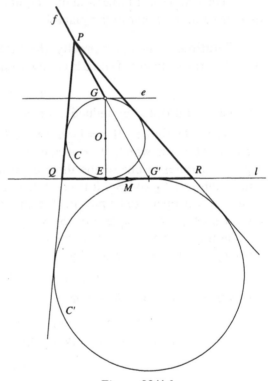

Magnify now C from P into the excircle C' of $\triangle PQR$ touching the side QR. This enlargement is mapping e into l and thus C' and l are touching each other at G', the image of G under this enlargement. As we know (see [19]), $QE = RG' = s - PR$ where s is the semiperimeter of the triangle PQR. ($PR \geq PQ$) can be assumed and hence the points E and G' are symmetric through M. This, of course, is including the case when the two points, E and G' happen to be identical.

Figure 92/4.1

It can be checked, on the diagram, that the configuration E, M, G', G and also the position of the line GG' depends on the given data only; the point P hence is incident to the straight line $G'G$ or, more precisely, it is on the opened ray f starting at G and separated from the circle C by e.

We now prove that this ray is the locus itself. Consider an arbitrary point P on f. Being on the opposite sides of the line e, both tangents from P to C

intersect e and also l; the latter at points Q and R respectively. By the construction C is the incircle of $\triangle PQR$. As we have already seen, the point G' where PG and l meet is the touching point of the excircle of the same triangle and thus $QE = G'R$. O and M are the midpoints of the sides GE and EG' respectively, M is hence halving QR, indeed.

The locus of the points P of the problem is the opened ray f.

1992/5. *Let S be a finite set of points in three-dimensional space. Let S_x, S_y, S_z be the sets consisting of the orthogonal projections of the points of S onto the yz-plane, xz-plane, xy-plane respectively. Prove that*

$$|S|^2 \leq |S_x||S_y||S_z|,$$

where $|A|$ denotes the number of points in the set A.

[The orthogonal projection of a point onto a plane is the foot of the perpendicular from the point to the plane.]

Solution. Denote, for brevity, the cardinalities of the sets S, S_x, S_y, S_z by n, a, b, c respectively. The task is now to prove

(1) $$n^2 \leq abc$$

and we shall proceed by induction on n.

(1) is clearly true if $n = 1 = a = b = c$. Let $n > 1$ and assume that (1) is true for any set S for which $|S| < n$ and let S be an n-element set. Consider now a plane, parallel to one of the three coordinate planes, which splits S into non empty subsets: S_1 and S_2. It is clear that one can always find such a plane, even if the set S happens to be planar. Assume that this separating plane is parallel to the xy-plane, for example. Let $|S_1| = n_1$, $|S_2| = n_2$; then, of course, $n_1 + n_2 = n$. Denote the number of points in the projections of the two sets, S_1 and S_2 on the yz-plane, zx-plane, xy-plane by

$$a_1, \; b_1, \; c_1, \qquad \text{and} \qquad a_2, \; b_2, \; c_2$$

respectively. We now clearly have

$$a_1 + a_2 = a, \qquad b_1 + b_2 = b, \qquad c_1 \leq c, \; c_2 \leq c.$$

Since both n_1 and n_2 are strictly less than n

$$n_1^2 \leq a_1 b_1 c_1 \qquad \text{and} \qquad n_2^2 \leq a_2 b_2 c_2$$

by the induction hypothesis. Hence

$$n^2 = (n_1 + n_2)^2 \leq \left(\sqrt{a_1 b_1 c_1} + \sqrt{a_2 b_2 c_2}\right)^2 \leq$$

$$\leq \left(\sqrt{a_1 b_1 c} + \sqrt{a_2 b_2 c}\right)^2 = c\left(\sqrt{a_1 b_1} + \sqrt{a_2 b_2}\right)^2.$$

Cauchy's inequality ([22]) now settles the issue since:

$$n^2 \leq c(a_1 + a_2)(b_1 + b_2) = abc,$$

and the proof is complete.

Remark. Equality might hold in (1) if, for example, the given points are all on a line parallel to one of the axes.

1992/6. *For each positive integer n, $S(n)$ is defined as the greatest integer such that for every positive integer $k \leq S(n)$, n^2 can be written as the sum of k positive squares.*

(a) *Prove that $S(n) \leq n^2 - 14$ for each $n \geq 4$.*

(b) *Find an integer n such that $S(n) = n^2 - 14$.*

(c) *Prove that there are infinitely many integers n such that $S(n) = n^2 - 14$.*

Solution. To get an impression about the flavour of $S(n)$ let's check how it works if $n = 13$. By definition $S(13)$ is the greatest positive integer for which $13^2 = 169$ can be written as the sum of $1, 2, 3, \ldots, S(13)$ squares. (By a square now we mean a positive number, zero is excluded.)It is easy to check, for example, that $S(13) \geq 12$, because $13^2 = 169$ can be written as the sum of $1, 2, \ldots,$ 12 squares as follows:

(1) $169 \overset{1}{=} 13^2 \overset{2}{=} 5^2 + 12^2 \overset{3}{=} 3^2 + 4^2 + 12^2 \overset{4}{=} 4^2 + 5^2 + 2 \cdot 8^2 \overset{5}{=} 5^2 + 4 \cdot 6^2 \overset{6}{=}$

$\overset{6}{=} 3^2 + 4^2 + 4 \cdot 6^2 \overset{7}{=} 5 \cdot 4^2 + 5^2 + 8^2 \overset{8}{=} 4 \cdot 3^2 + 5^2 + 3 \cdot 6^2 \overset{9}{=} 4^2 + 5 \cdot 3^2 + 3 \cdot 6^2 \overset{10}{=}$

$\overset{10}{=} 9 \cdot 4^2 + 5^2 \overset{11}{=} 3^2 + 10 \cdot 4^2 \overset{12}{=} 1^2 + 4 \cdot 2^2 + 5 \cdot 4^2 + 2 \cdot 6^2;$

The numbers upon the equality signs indicate the number of terms in the subsequent sum.

Let's turn now, one by one, to the solution of the three parts.

(a) It is enough to show that n^2 cannot be written as the sum of $n^2 - 13$ squares if $n \geq 4$. Assume the contrary: suppose that there exist $n^2 - 13$ positive integers such that

$$a_1^2 + a_2^2 + \ldots + a_{n^2-13}^2 = n^2.$$

Rearranging we get

(2) $$(a_1^2 - 1) + (a_2^2 - 1) + \ldots + (a_{n-13}^2 - 1) = 13.$$

The terms in the brackets are not negative and thus $a_i^2 - 1 \leq 13$. Hence $a_i^2 - 1$ is either 0 or 3 or 8. There is at most one 8 in (2) because $2 \cdot 8$ is already greater than 13. If there is no 8 at all then the *l.h.s.* is a multiple of 3, a contradiction. Similarly, if there is exactly one 8 then the *l.h.s.* is congruent to 2 modulo 3 while 13 is not. Thus (2) is impossible, indeed.

(b) We prove that $n = 13$ will do by showing that $S(13) = 13^2 - 14 = 155$. Having already seen, in part (a), that $S(13) \leq 13^2 - 14 = 155$ it is left to show that 169 can be written as the sum of $1, 2, 3, \ldots, 155$ squares. This means that

for each k between 1 and 155 there exist positive integers a_i such that
$$169 = a_1^2 + a_2^2 + \ldots + a_k^2.$$
As in (2) this sum is written as

(3) $\qquad\qquad (a_1^2 - 1) + (a_2^2 - 1) + \ldots + (a_k^2 - 1) = 169 - k.$

Since $a_i^2 - 1 < 169$ we can prepare the list of the possible values of $a_i^2 - 1$ again:

(4) $\qquad\qquad$ 0, 3, 8, 15, 24, 35, 48, 63, 80, 99, 120, 143, 168.

We start by $(a_1^2 - 1)$; its value is adjusted by a kind of greedy algorithm; it is set to be the highest element not exceeding $155 - k$ on the list above. The possible values of $a_1^2 - 1$ are arranged in the following array:

$a_1^2 - 1$	a_1	$169 - k$	k
0	1^2	$14 \leq 169 - k \leq 16$	$153 \leq k \leq 155$
3	2^2	$17 \leq 169 - k \leq 21$	$148 \leq k \leq 152$
8	3^2	$22 \leq 169 - k \leq 28$	$141 \leq k \leq 147$
15	4^2	$29 \leq 169 - k \leq 37$	$132 \leq k \leq 140$
24	5^2	$38 \leq 169 - k \leq 48$	$121 \leq k \leq 131$
35	6^2	$49 \leq 169 - k \leq 61$	$108 \leq k \leq 120$
48	7^2	$62 \leq 169 - k \leq 76$	$93 \leq k \leq 107$
63	8^2	$77 \leq 169 - k \leq 93$	$76 \leq k \leq 92$
80	9^2	$94 \leq 169 - k \leq 112$	$57 \leq k \leq 75$
99	10^2	$113 \leq 169 - k \leq 133$	$36 \leq k \leq 56$
120	11^2	$134 \leq 169 - k \leq 156$	$13 \leq k \leq 35$
143	12^2	$157 \leq 169 - k \leq 168$	$1 \leq k \leq 12$

Having set the value of a_1 the further a_i-s have to be chosen according to
$$\left(a_2^2 - 1\right) + \left(a_3^2 - 1\right) + \ldots + \left(a_k^2 - 1\right) = 169 - k - \left(a_1^2 - 1\right).$$

By the choice of a_1 we have $14 \leq 169 - k - \left(a_1^2 - 1\right) \leq 36$ and now this number has to be written as a sum whose terms are of the form $a^2 - 1$. There are various ways to do this. If, for example, $169 - k - \left(a_1^2 - 1\right)$ is a multiple of 3 then it can be written as the sum of at most twelve 3-s; if it is equal to $3l + 1$ then two 8-s and at most six 3-s are needed; finally, if it is of the form $3l + 2$ then one 8 and at most nine 3-s will do the job.

Thus $169 - k - \left(a_1^2 - 1\right)$ is decomposed into the sum of at most 12 numbers, each of the form $a^2 - 1$. If $k \geq 13$ then the remaining terms in (3) are simply zero, therefore 169 can also be expressed as the sum of 13, 14, ..., 155 positive squares; as for fewer square terms, well, this has already been verified in the introductory part. Hence 13 is indeed a number requested in part (b).

(c) For this we prove that if $n \geq 8$ and $S(n) = n^2 - 14$ then $S(2n) = (2n)^2 - 14$. The claim now follows by induction: $n = 13 \cdot 2^m$ has the desired property for every m.

For $m = 0$ that is $n = 13$ we have already proved in part (b) that $S(n) = n^2 - 14$. Assume now that $S(n) = n^2 - 14$. We show that $4n^2$ can be written as the sum of k squares for every k between 1 and $4n^2 - 14$; this, when combined with (a), already implies that

$$S(2n) = 4n^2 - 14.$$

Consider first

(5) $$1 \leq k \leq n^2 - 14.$$

By the induction hypothesis n^2 now can be written as the sum of k positive squares; multiplying each term by 4 yields a decomposition of $4n^2$.

Let k_1, k_2, k_3, k_4 be now positive integers not exceeding $n^2 - 14$. By the induction hypothesis n^2 can be expressed as the sum of k_1, k_2, k_3, k_4 squares respectively:

$$n^2 = a_1^2 + a_2^2 + \ldots + a_{k_1}^2, \qquad n^2 = b_1^2 + b_2^2 + \ldots + b_{k_2}^2,$$
$$n^2 = c_1^2 + c_2^2 + \ldots + c_{k_3}^2, \qquad n^2 = d_1^2 + d_2^2 + \ldots + d_{k_4}^2.$$

Adding these sums yields a decomposition of $4n^2$ into the sum of $k_1 + k_2 + k_3 + k_4$ squares.

$$4n^2 = a_1^2 + \ldots + a_{k_1}^2 + b_1^2 + \ldots + b_{k_2}^2 + c_1^2 + \ldots + c_{k_3}^2 + d_1^2 + \ldots + d_{k_4}^2.$$

As for $k_1 + k_2 + k_3 + k_4$, the number of terms clearly

(6) $$4 \leq k \leq 4\left(n^2 - 14\right) = 4n^2 - 56.$$

Finally write n^2 as the sum of 1, 2, ..., $n^2 - 14$ squares respectively and append $3n^2$ copies of 1^2 to each decomposition. This yields a square sum representation of $4n^2$ where k, the number of terms satisfies

(7) $$3n^2 + 1 \leq k \leq 4n^2 - 14.$$

Observe additionally that if $n \geq 8$ then $4 < n^2 - 14$ and also $3n^2 + 1 < 4n^2 - 56$, and thus every k satisfying

$$1 \leq k \leq 4n^2 - 14,$$

fits to one of the conditions (5), (6) or (7) and thus the corresponding decomposition works.

1993.

1993/1. *Let* $f(x) = x^n + 5x^{n-1} + 3$, *where* $n > 1$ *is an integer. Prove that* $f(x)$ *cannot be expressed as the product of two non-constant polynomials with integer coefficients.*

First solution. Assume, to the contrary that $f(x)$ can be factorized as $f(x) = g(x)h(x)$ where

$$g(x) = a_n x^n + a_{n-1} x^{n-1} + \ldots + a_1 x + a_0,$$
$$h(x) = b_n x^n + b_{n-1} x^{n-1} + \ldots + b_1 x + b_0;$$

the coefficients are integers. Comparing the coefficients of $f(x)$ and those of the product, $g(x)h(x)$ yields

(1) $a_0 b_0 = 3,$

(2) $a_1 b_0 + a_0 b_1 = 0,$

(3) $a_2 b_0 + a_1 b_1 + a_0 b_2 = 0,$

$$a_3 b_0 + a_2 b_1 + a_1 b_2 + a_0 b_3 = 0,$$

$$\vdots$$

$$a_{n-2} b_0 + a_{n-3} b_1 + a_{n-4} b_2 + \ldots + a_1 b_{n-3} + a_0 b_{n-2} = 0.$$

(1) implies that a_0 or b_0 is equal to 3 or -3; we may clearly assume that $a_0 = 3$ and thus $b_0 = 1$. Plugging these values to (2) yields that 3 divides a_1; hence a_2 also and so on: the coefficients $a_3, a_4, \ldots, a_{n-2}$ are all divisible by 3.

Since $h(x)$ is at least first degree, by condition, a_n has to be zero otherwise the degree of the product $g(x)h(x)$ would exceed that of $f(x)$. We show now that a_{n-1} cannot be zero. Indeed, the opposite implies that every coefficient of $g(x)$ is divisible by 3 and thus its integer values as well. Then, of course, this also holds for the product $f(x)$. But $f(2) = 7 \cdot 2^{n-1} + 3$ is not a multiple of 3, a contradiction.

Since $a_{n-1} \neq 0$ the degree of $g(x)$ is $(n-1)$ and thus $h(x)$ is a first degree, it can be written as $h(x) = b_1 x + 1$. Comparing once more the coefficients yields $a_{n-1} b_1 = 1$ that is $|b_1| = 1$, $h(1) = 0$ or 2, an even number anyway. On the other hand $f(1) = 9$ is odd, a contradiction and thus the proof is finished.

Second solution. Suppose again that there is a factorization $f(x) = g(x)h(x)$; g and h are at least first degree integer polynomials. We may clearly assume that the leading coefficient of $g(x)$ — and hence that of $h(x)$ — is equal

to 1. Thus
$$g(x) = x^k + a_{k-1}x^{k-1} + a_{k-2}x^{k-2} + \ldots + a_1 x + a_0.$$

Comparing again the coefficients of $f(x)$ and $g(x)h(x)$ the product of the constant terms, $g(0)h(0)$, is 3 that is $|g(0)||h(0)| = 1 \cdot 3$. Hence

(4) $$|a_0| = |g(0)| = 1$$

can clearly be assumed. $k = 1$ is not possible otherwise $g(x)$ would be equal to $x \pm 1$ with a root -1 or 1 but none of these numbers fit $f(x)$; hence

(5) $$k > 1.$$

Consider now the roots $\alpha_1, \alpha_2, \ldots, \alpha_k$ of $g(x)$. Then $g(x)$ can be factorized as

(6) $$g(x) = (x - \alpha_1)(x - \alpha_2)\ldots(x - \alpha_k).$$

In the general case these roots are complex and, of course, they also make $f(x)$ zero. (4) now implies that

(7) $$|g(0)| = |\alpha_1\alpha_2\ldots\alpha_k| = 1.$$

Since $f(\alpha_i)$ is also zero $(i = 1, 2, \ldots, k)$,
$$\alpha_i^n + 5\alpha_i^{n-1} = -3,$$

(8) $$\alpha_i^{n-1}(\alpha_i + 5) = -3.$$

Prepare now the product of the relations of the kind (8) for $i = 1, 2, \ldots, k$:
$$(\alpha_1\alpha_2\ldots\alpha_k)^{n-1}(\alpha_1 + 5)(\alpha_2 + 5)\ldots(\alpha_k + 5) = (-3)^k.$$

By (7), this implies
$$|(\alpha_1 + 5)(\alpha_2 + 5)\ldots(\alpha_k + 5)| = 3^k.$$

The product on the *l.h.s.*, by (6), is equal to $g(-5)$; thus
$$|g(-5)| = 3^k.$$

On the other hand
$$g(-5)h(-5) = f(-5) = (-5)^n + 5(-5)^{n-1} + 3 = (-5)^{n-1}(-5+5) + 3 = 3.$$

Hence $|3^k \cdot h(-5)| = 3$ that is $|3^{k-1} \cdot h(-5)| = 1$. We have seen, however, that $k > 1$ and thus the last equality cannot hold because $h(-5)$ is also an integer; the required factorization is hence impossible, indeed.

1993/2. *Let D be a point inside the acute-angled triangle ABC such that $\angle ABD = 90° + \angle ACB$ and $AC \cdot BD = AD \cdot BC$.*

(a) *Calculate the ratio $\dfrac{AB \cdot CD}{AC \cdot BD}$.*

(b) *Prove that the tangents at C to the circumcircles of ACD and BCD are perpendicular.*

First solution. It is worth noting that the two parts of the problem are just vaguely related; the second condition, for example, is not necessary for the second claim to hold.

Let $\angle CAD = \alpha'$, $\angle CBD = \beta'$; besides the usual notations let $AD = a'$, $BD = b'$ and $CD = c'$ (*Figure 1993/2.1*). Consider part (b) first.

(b) Summing the angles in $\triangle ABD$:

$$180° = \alpha - \alpha' + \beta - \beta' + \gamma + 90°,$$

(3) that is $\alpha' + \beta' = 90°.$

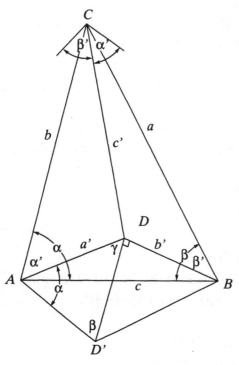

Figure 93/2.1

Draw a ray from C which makes an angle α' with DC opposite to A. By the theorem of inscribed angles this ray is touching the circumcircle of $\triangle ACD$ at C. Similarly, the ray making $\angle \beta'$ with DC at C, opposite to B is touching the circumcircle of $\triangle BCD$. The angle of these two tangents is $\alpha' + \beta'$ which, by (3), is equal to $90°$: the two tangents are thus perpendicular indeed; the second proposition is hence proved.

(a) Rotate and enlarge $\triangle ABC$ about A by α' with scale factor $\dfrac{a'}{b}$ mapping hence AC to AD. Denoting the image of D under this transformation by D' clearly $\angle ADD' = \gamma$ and $\angle DAD' = \alpha$.

With our notations (2) becomes

(4) $bb' = aa'.$

For the image DD' of CB clearly

$$DD' = a \cdot \frac{a'}{b} = \frac{bb'}{b} = b'; \text{ moreover } \angle BDD'\angle = 90°; \triangle BDD' \text{ is an isosceles right}$$

triangle and thus $BD' = b'\sqrt{2}$. Since $\angle DAD' = \alpha'$ $\angle BAD' = \alpha$. Hence $AB = c$ is

mapped into AD' and thus for the latter we have $AD' = \dfrac{ca'}{b}$.

Observe that $\triangle CAD$ and $\triangle BAD'$ are similar; indeed, they have an equal angle both at A and the respective ratios of the sides forming this angle are also

equal. The scale factor of similarity is $\dfrac{c}{b}$ and thus

$$BD' = \frac{c}{b} \cdot CD = \frac{cc'}{b} = b'\sqrt{2}, \quad \text{yielding} \quad \frac{cc'}{bb'} = \sqrt{2},$$

or

$$\frac{AB \cdot CD}{AC \cdot BD} = \sqrt{2},$$

the answer for the first question of the problem.

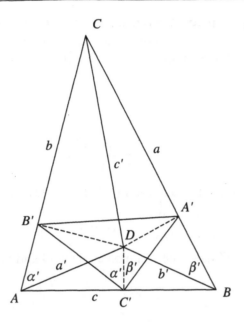

Figure 93/2.2

Second solution. (For part (a) only.) Drop perpendiculars from D to the sides AB, BC, CA; the feet are interior to the sides, by condition. Denote them by C', A', B' respectively. (See *Figures 1993/2.1* and *1993/2.2*) The quadrilateral $AC'DB'$ is clearly cyclic and thus, by the theorem of inscribed angles, $\angle DC'B' = \alpha'$ and, similarly, $\angle DC'A' = \beta'$. Hence, by (3)

$$B'C'A'\angle = \alpha' + \beta' = 90°,$$

$\triangle B'C'A'$ is right angled.

The diameter of the circumcircle of $AC'DB'$ is a', therefore

$$B'C' = a' \sin \alpha.$$

If, for simplicity, the circumradius of $\triangle ABC$ is $1/2$ then $a = \sin \alpha$ and

$$B'C' = aa'.$$

Similarly, $C'A' = bb'$ and $A'B' = cc'$. The second condition as it is written in (4) now implies

$$B'C' = C'A',$$

the right angled $\triangle B'C'A'$ is isosceles. Its hypotenuse is $A'B' = C'A'\sqrt{2}$ and thus, by the previous $cc' = bb'\sqrt{2}$,

$$\frac{cc'}{bb'} = \frac{AB \cdot CD}{AC \cdot BD} = \sqrt{2}.$$

Third solution. We apply *inversion*; this approach reveals a closer relation between the two parts and also the possible origin of the whole problem.

The following property of inversion will be used: if the images of the points X and Y under the inversion of pole O and radius R are X' and Y' respectively then — unless X, Y and O are collinear — $\triangle OXY$ and $\triangle OY'X'$ are similar and thus

$$\angle OXY = \angle OY'X' \quad \text{and} \quad \angle OYX = \angle OX'Y',$$

moreover

(5)
$$X'Y' = \frac{R^2 \cdot XY}{OX \cdot OY}.$$

Let's see the proof. By definition $OX \cdot OX' = OY \cdot OY' = R^2$ therefore

$$\frac{OX}{OY} = \frac{OY'}{OX'}. \quad \text{(Figure 1993/2.3)}$$

$\triangle OXY$ and $\triangle OY'X'$ have one common angle at O and the ratios of the respec-

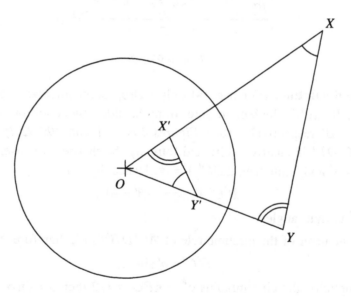

Figure 93/2.3

tive sides forming this angle are also equal; the two triangles are hence similar, indeed, therefore

$$\frac{X'Y'}{XY} = \frac{OX'}{OY}, \quad \text{that is} \quad X'Y' = \frac{OX' \cdot XY}{OY}.$$

Substituting

$$OX' = \frac{R^2}{OX}$$

yields (5). Observe that (5), of course, holds even if the points O, X and Y are collinear.

Apply now inversion of pole D and radius $R = \sqrt{a'b'c'}$ to the points A, B, C in *Figure 1993/2.1*. By our previous remark these points transform into the triple A', B', C' such that

$$A'B' = cc', \qquad B'C' = aa', \qquad C'A' = bb'. \quad (Figure\ 1993/2.4)$$

With the notations of *Figure 1993/2.1* the previous results become

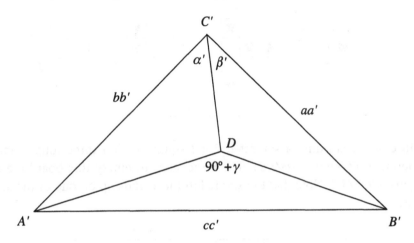

Figure 93/2.4

$$\alpha' = \angle DAC = \angle DC'A' \qquad \text{and} \qquad \beta' = \angle DBC = \angle DC'B'.$$

Hence $\angle A'C'B' = \alpha' + \beta'$ and in the previous solutions we have already seen that these angles are complementary. $\triangle A'C'B'$ is hence right angled and isosceles and thus

$$\frac{cc'}{bb'} = \frac{A'B'}{C'A'} = \sqrt{2}.$$

The circumcircles of ACD and BCD are mapped to the straight lines $A'C'$ and $B'C'$ respectively. Since inversion is preserving angles and $A'C'$ and $B'C'$ are perpendicular, the two circles above are also orthogonal; their respective tangents at C make a right angle, indeed; proposition (b) is hence proved.

1993/3. *On an infinite chessboard a game is played as follows. At the start n^2 pieces are arranged in an $n \times n$ block of adjoining squares, one piece on each square. A move in the game is a jump in a horizontal or vertical direction over an adjacent occupied square to an unoccupied square immediately beyond. The piece which has been jumped over is removed. Find those values of n for which the game can end with only one piece remaining on the board.*

Solution. We show that the object of the game can be achieved if and only if 3 does not divide n. We shall proceed by induction and, getting started, two particular cases are going to be checked.

In case a) consider a board where there are two squares adjoined to the extreme right field of a 3×1 rectangle. There are four pieces altogether on the board and one of the adjoined squares is unoccupied. *Figure 1993/3.1* shows how this board can be emptied. The pieces are shown as spots and the arrows indicate the three moves. As a result, a 3×1 rectangle can always be emptied if there are two adjacent squares such that just one of them is occupied.

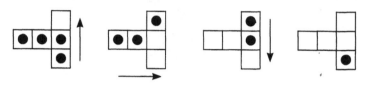

Figure 93/3.1

In case b) consider a saturated 2×2 square with a three long empty row attached to it (*Figure 1993/3.2*). The three steps to empty this board are shown in *Figure 1993/3.2*. Note that the single left piece ends up on one of the attached squares.

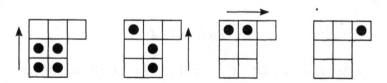

Figure 93/3.2

After this preliminary solitaire let's turn to the induction proof; we show that if $n = 3k + 1$ or $n = 3k + 2$ then one can arrive to a single piece on the board.

This is obvious if $n = 1$ and if $n = 2$ then this is case b). Let $n > 1$ and assume that the task can be done on any board whose dimension is not divisible by 3 and less than n.

If $n = 3k + 1$ then consider a framing pattern of 3×1 rectangles of the $n \times n$ board as it is shown in *Figure 1993/3.3*. Since the externally adjacent squares are all empty, the frame can be cleaned if proceeding piecewise as in case a). Having finished we are left with a $3k + 1 - 2 = 3(k - 1) + 2$ saturated board and the rest is induction.

The story is essentially the same if $n = 3k + 2$. The frame now has two layers (*Figure 1993/3.4*) and beginning from the outside both layers can be cleaned up following again the procedure in a). The full part of the board is of side $3k + 2 - - 4 = 3(k - 1) + 1$ so, by the induction hypothesis, we are done, again.

Finally, we prove that the solitaire has no solution if the side of the board is $3k$. Divide the *infinite* board into 3×3 squares and label the fields as it is shown below.

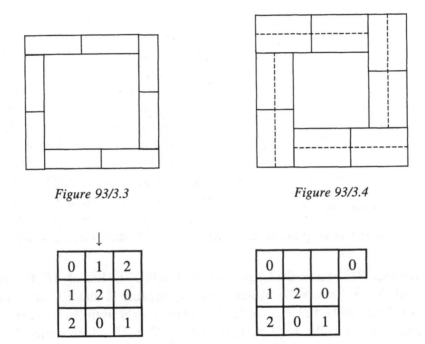

Figure 93/3.3   Figure 93/3.4

Consider now a $3k \times 3k$ occupied block and mark each piece with the label of the currently occupied square. There are two squares emptied during any move and another one so far empty gets occupied. Jumping to the right, for example, from the indicated square in the figure the two pieces removed were labelled by 1 and 2 respectively while the apparently new piece pops up with label 0. Thus, as a byproduct of every move, the number of labelled pieces is changing by 1 for each label; there are two of them, like 1 and 2 in the example, whose total is decreasing and for the third one it is increasing. Initially there are the same number of pieces labelled by 0, 1 and 2 respectively (n is now a multiple of 3) and thus, as long as we are playing, the *parity* of the number of pieces is the same for the respective labels.

This is the heart of the issue; now it is clear that we cannot end up with a single piece: the label of its current field would occur but once while the other two labels would be represented zero times, thus breaking the parity invariance.

1993/4. *For three points P, Q, R in the plane define* $m(PQR)$ *as the minimum length of the three altitudes of the triangle PQR (or zero if the points are collinear). Prove that for any points A, B, C, X*

$$m(ABC) \le m(ABX) + m(AXC) + m(XBC).$$

Solution. In what follows by 'the longest one' in a system of distances we mean one of the longest distances in the system.

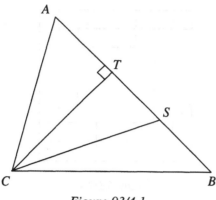

Figure 93/4.1

The heart of the issue is the following *Lemma*. If S is an interior point of the segment AB then $m(ABC) \geq m(ASC)$.

To prove this it is clearly enough to show that $m(ABC)$ is not shorter than some altitude of $\triangle ASC$ because then it obviously cannot be shorter than the smallest altitude of the triangle.

Assume first that AB is the longest side of $\triangle ABC$ (*Figure 1993/4.1*). Since there is the shortest height perpendicular to the longest side now it is the height CT; this, at the same time, is the height of $\triangle ASC$ and thus the claim is now true.

Let side BC be now the longest one in $\triangle ABC$. The line of AR, the shortest height of $\triangle ABC$ meets CS at some point Q because A and BC are separated by CS (*Figure 1993/4.2*). Denote the foot of the height from A of $\triangle ASC$ by Z. From the right triangle AZQ we have $AZ < AQ < AR$ which settles the claim in this case.

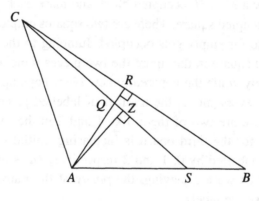

Figure 93/4.2

Finally, if the longest side of $\triangle ABC$ is AC then the smallest height is BT and for the perpendicular SR from S to AC clearly $SR < BT$ (*Figure 1993/4.3*). Being so

$$m(ASC) = SR < BT = m(ABC).$$

Turning to the actual problem the argument depends on the position of the point X relative to $\triangle ABC$. There are three possibilities:

1. X belongs to the interior or the boundary;

2. X belongs to the so called U-*regions* that contain the excircles of the triangle;

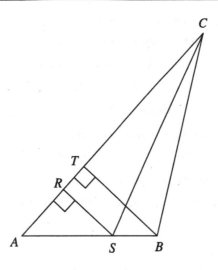

Figure 93/4.3

or

3. X is in the so called *V-regions* formed by the angles vertical to the angles of ABC.

1. X is now contained by $\triangle ABC$. Assume also that with standard notations $a \geq b \geq c$ (*Figure 1993/4.4*). Expressing the area of $\triangle ABC$ in two different ways:

(1) $2[ABC] = a \cdot m(ABC) = a_1 \cdot m(ABX) + a_2 \cdot m(AXC) + a_3 \cdot m(XBC),$

where a_1, a_2 and a_3 denote the longest sides of the triangles ABX, AXC, ABC

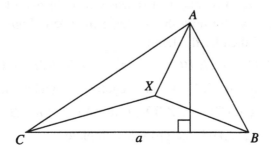

Figure 93/4.4

respectively. Since the longest distance contained in a triangle cannot exceed the longest side of the same triangle we have $a_1 \leq a$, $a_2 \leq a$, $a_3 \leq a$ and thus (1) implies

$$a \cdot m(ABC) \leq a \cdot m(ABX) + a \cdot m(AXC) + a \cdot m(XBC).$$

Dividing through by a the claim follows.

2. X is now in one of the U-regions, say the one opposite to A. (*Figure 1993/4.5*) Denote the intersection of AX and BC by H. By our lemma

(2) $\qquad m(ABH) \leq m(ABX) \qquad$ and $\qquad m(AHC) \leq m(AXC).$

We reuse the results proved in part 1. For $\triangle ABC$ and the point H:

(3) $\qquad\qquad m(ABC) \leq m(ABH) + m(AHC) + m(HBC).$

Now $m(HBC) = 0$ and thus (2) and (3) yield

$$m(ABC) \leq m(ABX) + m(AXC) + 0.$$

Since $m(XBC) \geq 0$ the estimation can be extended and we get the claim again:

$$m(ABC) \leq m(ABX) + m(AXB) + m(XBC).$$

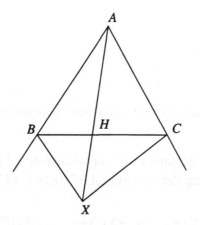 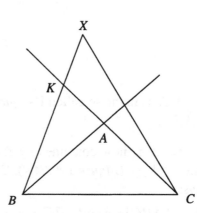

Figure 93/4.5   *Figure 93/4.6*

3. The last case to be checked is when X is in one of the V-regions, say the one at A, the rays forming the V-shape included (*Figure 1993/4.6*) If BX and AC meet at K then by the lemma

(4) $\qquad m(ABC) \leq m(KBC) \qquad$ and $\qquad m(KBC) \leq m(XBC).$

Since $m(ABX)$ and $m(AXC)$ are not negative (4) yields the claim

$$m(ABC) \leq m(ABX) + m(AXC) + m(XBC);$$

each case has been checked, the proof is complete.

1993/5. *Does there exist a function f from the positive integers to the positive integers such that $f(1) = 2$, $f(f(n)) = f(n) + n$ for all n, and $f(n) < f(n+1)$ for all n?*

First solution. Denote the positive root of the equation

(4) $\qquad\qquad\qquad y^2 - y - 1 = 0$

by a; it is equal to $\dfrac{1}{2}(1 + \sqrt{5}) = 1,16180\ldots$.

Define now the sequences

$$g(n) = an, \qquad f(n) = \left[g(n) + \frac{1}{2}\right] = \left[an + \frac{1}{2}\right].$$

We show that $f(n)$ satisfies the conditions and hence the answer for the question of the problem is affirmative. By (4)

(5) $$g(g(n)) - g(n) - n = 0.$$

Since $f(1) = [1,6180\ldots + 0,5] = 2$, (1) holds; moreover, $a > 1$ implies

$$f(n+1) = \left[an + \frac{1}{2} + a\right] \geq \left[an + \frac{1}{2} + 1\right] = \left[an + \frac{1}{2}\right] + 1 = f(n) + 1 > f(n)$$

and thus (3) is also satisfied.

Writing it as $n + f(n) - f(f(n)) = 0$ we follow a tricky way to prove (2): since the values of f are whole numbers it is enough to show that

$$|n + f(n) - f(f(n))| < 1, \qquad \text{for every } n \in \mathbf{N}.$$

We note first that by the irrationality of $g(n)$

(6) $$|f(n) - g(n)| < \frac{1}{2}.$$

Now clearly $g(n_1) - g(n_2) = a(n_1 - n_2)$ $(n_1 \in \mathbf{N},\ n_2 \in \mathbf{N})$ and (5) can be rearranged as $n = g(g(n)) - g(n)$; comparing these pieces:

$$n + f(n) - f(f(n)) = g(g(n)) - g(f(n)) + g(f(n)) - f(f(n)) + f(n) - g(n) =$$
$$= a(g(n) - f(n)) - (g(n) - f(n)) + g(f(n)) - f(f(n)) =$$
$$= (a - 1)(g(n) - f(n)) + g(f(n)) - f(f(n)).$$

Hence

$$|n + f(n) - f(f(n))| \leq (a - 1)|g(n) - f(n)| + |g(f(n)) - f(f(n))| <$$
$$< (a - 1)\frac{1}{2} + \frac{1}{2} = \frac{a}{2} < 1,$$

indeed, the proof is hence complete.

Second solution. In the solution we shall use the *Fibonacci-sequence* defined by the recurrence

$$u_1 = 1, \qquad u_2 = 2, \qquad u_{n+1} = u_n + u_{n-1} \quad (n \geq 2).$$

The initial terms of this increasing sequence are

$$u_1 = 1, \quad u_2 = 2, \quad u_3 = 3, \quad u_4 = 5, \quad u_5 = 8, \quad u_6 = 13, \quad u_7 = 21, \quad \ldots$$

the first few *Fibonacci numbers*. First we provide an algorithm which uniquely splits every positive integer into the sum of Fibonacci numbers such that there are no consecutive terms of the sequence in the representation. Single term sums are also accepted here. The proof goes by induction. For the first few positive integers this can be checked with bare hands. Let n be greater than 1 and assume that the claim holds for every positive integer less than n.

If $n = u_k$ then this is already the desired representation. If n itself is not a Fibonacci number then consider its Fibonacci neighbours, u_k and u_{k+1}, that is

(7) $$u_k < n < u_{k+1}.$$

Hence, by $n < u_{k+1} = u_k + u_{k-1} < 2u_k$

$$0 < n - u_k < u_k < n.$$

By the induction hypothesis $n - u_k$ as a positive integer less than n can be written as

$$n - u_k = u_{i_1} + u_{i_2} + \ldots + u_{i_s}$$

and thus

(8) $$n = u_{i_1} + u_{i_2} + \ldots + u_{i_s} + u_k,$$

where u_{i_s} and u_k are not consecutive Fibonacci numbers; the opposite, by the defining recurrence, would imply

$$n = u_{i_1} + u_{i_2} + \ldots + u_{k+1}$$

contradicting to (7). The induction is finished, the required representation does exist for every n.

Next we prove that this "Fibonacci representation" of positive numbers is unique. The very same induction works similarly and all we need to show is that u_k, as defined above, has to be present in any Fibonacci sum for n. Assume the contrary and consider a sum

$$n = u_{i_1} + u_{i_2} + \ldots + u_{i_s} + u_{i_{s+1}}$$

which does not include u_k. Thus $u_{i_{s+1}} < u_k$ and this implies the stronger inequality $u_{i_{s+1}} \leq u_{k-1}$.

Let k be even first; $k - 1$ is then odd and since there are no consecutive Fibonacci numbers in the sum, clearly

$$n = u_{i_1} + u_{i_2} + \ldots + u_{i_{s+1}} < (1 + u_1) + u_3 + u_5 + \ldots + u_{k-1} =$$
$$= (u_2 + u_3) + u_5 + \ldots + u_{k-1} = (u_4 + u_5) + \ldots + u_{k-1} = u_k \leq n,$$

a contradiction. Similarly, if k is odd that is $k - 1$ is even and thus

$$n = u_{i_1} + u_{i_2} + \ldots + u_{i_{s+1}} < (u_1 + u_2) + u_4 + u_6 + \ldots + u_{k-1} =$$
$$= (u_3 + u_4) + u_6 + \ldots + u_{k-1} = u_k \leq n,$$

a contradiction again; the decomposition in (8) is unique, indeed.

We can now define the function f in terms of this decomposition. If n is written as a Fibonacci sum in (8) then let $f(n)$ be

(9) $$f(n) = u_{i_1+1} + u_{i_2+1} + \ldots + u_{i_s+1} + u_{k+1}.$$

We have to show, of course, that this well defined function satisfies the conditions (1)–(3).

$$f(1) = f(u_1) = u_2 = 2;$$

that's about (1).

$$f(f(n)) = f\left(u_{i_1+1} + u_{i_2+1} + \ldots + u_{i_s+1} + u_{k+1}\right) =$$

$$= u_{i_1+2} + u_{i_2+2} + \ldots + u_{i_s+1} + u_{k+2} =$$

$$= \left(u_{i_1+1} + u_{i_2+1} + \ldots + u_{i_s+1} + u_{k+1} \right) + \left(u_{i_1} + u_{i_2} + \ldots + u_{i_s} + u_k \right) =$$

$$= f(n) + n,$$

so (2) is also settled.

We prove (3) by induction.

$$f(1) = f(u_1) = 2 < f(u_2) = u_3 = 3,$$

(3) thus holds for the initial value. Let $n > 1$ and assume that (3) holds for every positive integer less than n and let $u_k \leq n < u_{k+1}$ for some k.

We distinguish two cases:

A) $n + 1 < u_{k+1}$;

B) $n + 1 = u_{k+1}$.

A) The induction hypothesis now also implies that if $n' < n'' < n$ are arbitrary positive integers then

(10) $$f(n') < f(n'').$$

Indeed, applying (3) for the numbers all less that n yields the following chain of inequalities:

$$f(n') < f(n'+1) < f(n'+2) < \ldots < f(n'').$$

We know that both n and $n+1$ can be written as Fibonacci sums:

(11) $$n = u_{i_1} + u_{i_2} + \ldots + u_{i_s} + u_k = n' + u_k, \qquad (n' < n)$$

$$n + 1 = u_{j_1} + u_{j_2} + \ldots + u_{j_r} + u_k = n'' + u_k \qquad (n'' < n).$$

We have to show that $f(n) < f(n+1)$ that is

(12) $$u_{i_1+1} + u_{i_2+1} + \ldots + u_{i_s+1} + u_{k+1} < u_{j_1+1} + u_{j_2+1} + \ldots + u_{j_r+1} + u_{k+1},$$

or

$$u_{i_1+1} + u_{i_2+1} + \ldots + u_{i_s+1} < u_{j_1+1} + u_{j_2+1} + \ldots + u_{j_r+1}.$$

By the definition of f this is nothing else but

$$f(n') < f(n''),$$

and this follows, as we have seen, from the induction hypothesis.

B) With the notation in (11) $f(n) < f(n+1)$ is now

(13) $$u_{i_1+1} + u_{i_2+1} + \ldots + u_{i_s+1} + u_{k+1} < u_{k+2} = u_k + u_{k+1}$$

that is

$$u_{i_1+1} + u_{i_2+1} + \ldots + u_{i_s+1} < u_k.$$

By the definition of f this last inequality is nothing else but

$$f(n') < f(u_{k-1});$$

since $n' < n$ and $u_{k-1} < n$, the induction hypothesis yields the claim again.

We note that if $n = u_k$ then the sum $u_{i_1} + \ldots + u_{i_s}$ is replaced by zero and thus (12) and (13) obviously hold.

1993/6. *There are $n > 1$ lamps L_0, L_1, ..., L_{n-1} in a circle. We use L_{n+k} to mean L_k. A lamp is at all times either on or off. Perform steps s_0, s_1,... as follows: at step s_i, if L_{i-1} is lit, then switch L_i from on to off or vice versa, otherwise do nothing. Show that:*

(a) *There is a positive integer $M(n)$ such that after $M(n)$ steps all the lamps are on again;*

(b) *If $n = 2^k$, then we can take $M(n) = n^2 - 1$.*

(c) *If $n = 2^k + 1$ then we can take $M(n) = n^2 - n + 1$.*

Solution. (a) The following chart contains the instructions of the problem

	L_{j-1}	L_j	L_{j-1}	L_j	L_{j-1}	L_j	L_{j-1}	L_j
before S_j:	ON	ON	ON	OFF	OFF	ON	OFF	OFF
after S_j:	ON	OFF	ON	ON	OFF	ON	OFF	OFF

From the chart it is clear that the state after S_j uniquely determines the state before S_j. On the other hand there are but finitely many states of this n-lamp system and thus some state has to occur more than once. This repeating state, as we noted, uniquely determines the previous states, one by one, and thus, after a certain number of steps — this is the number $M(n)$ of the problem — each lamp will be ON, again.

It can be checked from the chart that if $a_k \equiv 1 \pmod 2$ then the kth lamp is ON after the $(k - n)$th step and it is OFF if $a_k \equiv 0 \pmod 2$. Remember that the lamps are numbered mod n.

In what follows we shall make use of the sequence a_0, a_1, ... defined as

(1) $\qquad a_0 = a_1 = \ldots = a_{n-1} = 1, \qquad a_k = a_{k-1} + a_{k-n} \qquad (k \geq n)$.

The *characteristic equation* of recurrence (1) is

$$x^k = x^{k-1} + x^{k-n},$$

which can also be written as

(2) $\qquad\qquad\qquad x^n - x^{n-1} - 1 = 0.$

This equation has no multiple roots (see the *Note*) and thus, as it is well known from the theory of *linear recurrences*, with given coefficients c_i

(3) $\qquad\qquad\qquad a_k = c_1 \alpha_1^k + c_2 \alpha_2^k + \ldots + c_n \alpha_n^k,$

where $\alpha_1, \alpha_2, \ldots, \alpha_n$ are the roots of equation (2).

(b) Turning to part (b) let n be some positive integral power of 2 and α a root of (2) that is $\alpha^n = \alpha^{n-1} + 1$. Consider now $\alpha^{n^2 + k}$. Clearly

$$\alpha^{n^2+k} = \alpha^k \cdot \alpha^{n^2} = \alpha^k (\alpha^n)^n = \alpha^k (\alpha^{n-1} + 1)^n =$$

$$= \alpha^k \left(\alpha^{n(n-1)} + \binom{n}{1} \alpha^{(n-1)(n-1)} + \ldots + \binom{n}{n-1} \alpha^{n-1} + 1 \right),$$

on the other hand

$$\alpha^{n^2+k} = \alpha^k \cdot \alpha^{n^2} = \alpha^k \cdot \alpha^{n^2-n} \cdot \alpha^n = \alpha^k \cdot \alpha^{n^2-n}(\alpha^{n-1}+1) = \alpha^k\left(\alpha^{n^2-1} + \alpha^{n(n-1)}\right).$$

Subtracting these two equalities

$$0 = \alpha^k \left(\binom{n}{1} \alpha^{(n-1)^2} + \ldots + \binom{n}{n-1} \alpha^{n-1} + 1 - \alpha^{n^2-1} \right).$$

Hence, by (3)

$$\binom{n}{1} a_{(n-1)^2+k} + \ldots + \binom{n}{n-1} a_{n-1+k} + a_k - a_{n^2-1+k} =$$

$$= \binom{n}{1} \sum_{i=1}^{n} c_i \alpha_i^{(n-1)^2+k} + \ldots + \binom{n}{n-1} \sum_{i=1}^{n} c_i \alpha_i^{n-1+k} + \sum_{i=1}^{n} c_i \alpha_i^{k} - \sum_{i=1}^{n} c_i \alpha_i^{n^2-1+k} =$$

$$= \sum_{i=1}^{n} c_i \left(\binom{n}{1} \alpha_i^{(n-1)^2+k} + \ldots + \binom{n}{n-1} \alpha_i^{n-1+k} + \alpha_i^{k} - \alpha_i^{n^2-1+k} \right) = \sum_{i=1}^{n} c_i \cdot 0 = 0.$$

Since n is a power of 2 the binomial coefficients in the above expressions are all even and thus, by the previous result,

$$0 = \binom{n}{1} a_{(n-1)^2+k} + \ldots + \binom{n}{n-1} a_{n-1+k} + a_k - a_{n^2-1+k} \equiv$$

$$\equiv a_k - a_{n^2-1+k} \quad (\text{mod } 2) \quad \text{that is}$$

$$a_k \equiv a_{n^2-1+k} \quad (\text{mod } 2).$$

Since $a_k = 1$ if $k = 0, 1, \ldots, n-1$-re $a_k = 1$

$$a_{n^2} \equiv a_{n^2+1} \equiv \ldots + a_{n^2+n-1} \equiv 1 \quad (\text{mod } 2),$$

the desired result.

(c) Now it is $(n-1)$'s turn to be a power of 2; using the notations and also the results of part (b) consider α^{n^2-n+k}, again. First of all

$$\alpha^{n^2-n+k} = \alpha^k \cdot \alpha^{n^2-n} = \alpha^k (\alpha^n)^{n-1} = \alpha^k \left(\alpha^{n-1} + 1 \right)^{n-1} =$$

$$= \alpha^k \left(\alpha^{(n-1)^2} + \binom{n-1}{1} \alpha^{(n-2)(n-1)} + \ldots + \binom{n-1}{n-2} \alpha^{n-1} + 1 \right).$$

On the other hand

$$\alpha^{n^2-n+k} = \alpha^k \cdot \alpha^{n^2-n} = \alpha^k \cdot \alpha^{(n-1)^2} \cdot \alpha^{n-1} = \alpha^k \cdot \alpha^{(n-1)^2} (\alpha^n - 1) =$$

$$= \alpha^k \left(\alpha^{n^2-n+1} - \alpha^{(n-1)^2} \right).$$

Subtracting these equalities

$$0 = \alpha^k \left(2\alpha^{(n-1)^2} + \binom{n-1}{1} \alpha^{(n-2)(n-1)} + \ldots + \binom{n-1}{n-2} \alpha^{n-1} + 1 - \alpha^{n^2-n+1} \right).$$

Proceeding now as in part (b)

$$0 = 2a_{(n-1)^2+k} + \binom{n-1}{1} a_{(n-2)(n-1)+k} + \ldots + \binom{n-1}{n-2} a_{n-1+k} + a_k - a_{n^2-n+1+k}.$$

The binomial coefficients are even again and thus

$$a_k \equiv a_{n^2-n+1+k} \pmod 2 \qquad \text{for every } k.$$

Since $a_1 = \ldots = a_n = 1$

$$a_{n^2-n+2} \equiv a_{n^2-n+3} \equiv \ldots \equiv a_{n^2+1} \equiv 1 \pmod 2,$$

and this was to be proved.

Remark. We have used that the polynomial

$$x^n - x^{n-1} - 1 = 0$$

has no multiple roots. Indeed, multiple roots also fit $nx^{n-1} - (n-1)x^{n-2} = x^{n-2}(nx - (n-1))$, the derivative polynomial; the roots of the latter are obviously 0 and $\dfrac{n-1}{n}$ and none of these do satisfy the original polynomial.

1994.

1994/1. *Let m and n be positive integers. Let a_1, a_2, ..., a_m be distinct elements of $\{1, 2, \ldots, n\}$ such that whenever $a_i + a_j \leq n$ for some i, j (possibly the same) we have $a_i + a_j = a_k$ for some k. Prove that*

$$\frac{a_1 + a_2 + \ldots + a_m}{m} \geq \frac{n+1}{2}.$$

Solution. Denote the set $\{a_1, a_2, \ldots, a_m\}$ by A; we may assume that the elements are in decreasing order that is $a_1 > a_2 > \ldots > a_m$. Assign, to any a_i element of A, the element a_{m-i+1} as its pair. This matching is symmetric, the pair of a_{m-i+1} is a_i and if m is odd then the median, $a_{\frac{m+1}{2}}$ is equal to its own pair.

First we show that the sum of any pair is at least $n+1$. If, to the contrary, $a_i + a_{m-i+1} < n+1$ for some $(1 \leq i \leq m)$ then the ordering of A yields

(1) $$a_i < a_i + a_m < a_i + a_{m-1} < \ldots < a_i + a_{m-i+1} \leq n.$$

Hence, by the condition, the i sums in (1) produce but elements of A, moreover, all of them are greater than a_i, a contradiction since there are only $i - 1$ elements in A beyond a_i. Therefore

$$a_1 + a_m \geq n + 1,$$

$$a_2 + a_{m-1} \geq n + 1,$$

$$\vdots$$

$$a_m + a_1 \geq n + 1.$$

Summing these inequalities yields the claim:

$$2(a_1 + a_2 + \ldots + a_m) \geq m(n+1).$$

1994/2. *ABC is an isosceles triangle with $AB = AC$. M is the midpoint of BC and O is the point on the line AM such that OB is perpendicular to AB. Q is an arbitrary point on BC different from B and C. E lies on the line AB and F lies on the line AC such that E, Q and F are distinct and collinear. Prove that OQ is perpendicular to EF if and only if $QE = QF$.*

Solution. Everything is obvious if $M = Q$; assume hence that $M \neq Q$. AM is on the axis of the triangle and thus O is diametrically opposite to A on the circumcircle (*Figure 1994/2.1*); besides $\angle OBC = \angle OCB$.

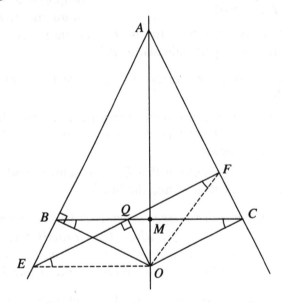

Figure 94/2.1

Assume first that OQ and EF are perpendicular. The quadrilaterals $OEBQ$ and $OCFQ$ are then cyclic because $\angle OBE = \angle OQE = \angle OCF = \angle OQF = 90°$. Hence, as inscribed angles

$$\angle OBQ = \angle OBC = \angle OEQ,$$

$$\angle OCQ = \angle OCB = \angle OFQ.$$

The right triangles OEQ and OFQ are thus congruent so $QE = QF$, indeed.

The converse is proved by the method of *reductio ad absurdum:* assume that

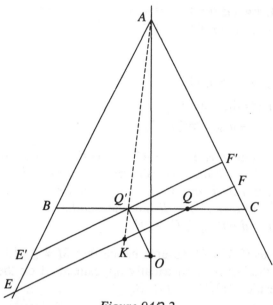

Figure 94/2.2

Q bisects EF but OQ and EF are still not perpendicular (*Figure 1994/2.2*). Draw perpendicular from O to EF; this line cuts BC at Q'. We need also the perpendicular to OQ' at Q'; it intersects the lines AB and AC at E' and F' respectively. By the first part of the proof Q'E' = Q'F' and now, by our indirect assumption the points Q and Q' are definitely different.

Consider now the median through Q' of △AE'F'. It cuts EF at the internal point K that bisects EF. This, however, is impossible, since Q is also the midpoint of EF and it is obviously different from K since the latter is not incident to BC. The contradiction proves that OQ is indeed perpendicular to EF.

Remark. In terms of the theory of conic sections the problem can be rephrased as follows:

If AB and AC are tangents to the parabola, moreover, they are symmetric with respect to its axis then the tangent at the vertex bisects the part cut by AB and AC of any other tangent.

The reason of this symmetry is that EF is touching another parabola whose focus is O and it touches at its vertex the line BC.

1994/3. *For any positive integer k, let $f(k)$ be the number of elements in the set $A_k = \{k+1, k+2, \ldots, 2k\}$ which have exactly three 1s when written in base 2. Prove that for each positive integer m, there is at least one k with $f(k) = m$ and determine all m for which there is exactly one k.*

Solution. Call a positive integer 'good', for brevity, if its binary form contains exactly three 1s; that is, it is the sum of three different powers of 2. Observe that k is good if and only if $2k$ is also good.

(a) Compare now the number of good elements in the sets A_k and A_{k+1}.

$$A_k = \{k+1, \; k+2, \; k+3, \; \ldots, \; 2k-1, \; 2k, \qquad\qquad\quad \},$$
$$A_{k+1} = \{ \qquad k+2, \; k+3, \; \ldots, \; 2k-1, \; 2k, \; 2k+1, \; 2k+2\}.$$

These sets differ in the elements $k+1$, and $2k+1$, $2k+2$ only. Since $k+1$ and $2k+2$ are equally good or not, $2k+1$ is responsible for any difference between

$f(k)$ and $f(k+1)$. Thus

$$f(k+1) - f(k) = 0, \qquad \text{if } 2k+1 \text{ is not good,}$$
$$f(k+1) - f(k) = 1, \qquad \text{if } 2k+1 \text{ is good.}$$

Increasing k by 1 either leaves $f(k)$ unchanged or it also goes up by 1 and the latter happens if and only if $2k+1$ is good.

Let us calculate now the value of $f(2^n)$. Since 2^{n+1} is not good there are the same number of good elements in the intervals $[1, 2^{n+1}]$ and $[1, 2^{n+1} - 1]$. The binary form of these good numbers contains $n+1$ digits (leading zeros are certainly possible) and three of them are equal to 1; there are $\binom{n+1}{3}$ such numbers. Similarly, $f(2^n) = \binom{n}{3}$. Thus the set

$$A_{2^n} = \{2^n + 1,\ 2^n + 2,\ \ldots,\ 2^{n+1}\} \qquad \text{contains} \quad f(2^n) = \binom{n+1}{3} - \binom{n}{3} = \binom{n}{2}$$

good numbers. Since $f(4) = f(2^2) = 1$ and the counting above implies that f assumes arbitrarily large values jumping at most 1 at a time, it follows that f admits every positive integer.

(b) Given the value of m the equation $f(k) = m$ has a single solution if f is strictly increasing at k that is

$$f(k+1) - f(k) = f(k) - f(k-1) = 1.$$

For this to happen both $2k+1$ and $2k-1$ have to be good as we have seen above. If $2k-1$ is good then both its first and last binary digits are 1. We show that the third unary digit is next to the last one. If not then adding 2 (10_2) to $2k-1$, the result, $2k+1$ would contain four 1s and thus it were not good. Thus

$$2k - 1 = 2^a + 2 + 1, \qquad (a \geq 3)$$

therefore

$$k = 2^{a-1} + 2. \quad \text{With} \quad n = a - 1 \quad \text{this is} \quad k = 2^n + 2 \qquad (n \geq 2).$$

The converse is obvious, if $k = 2^n + 2$ then $2k+1$ and $2k-1$ are both good.

We are left to determine $f(2^n + 2)$, the number of good elements in the set

$$A_{2^n+2} = \{2^n + 3,\ 2^n + 4,\ \ldots,\ 2^{n+1} + 4\}.$$

Since $2^{n+1} + 3$ is good while there is no good one among the numbers

$$2^n,\ 2^n + 1,\ 2^n + 2,\ 2^{n+1},\ 2^{n+1} + 1,\ 2^{n+1} + 2,\ 2^{n+1} + 4$$

there is one more good number in the set A_{2^n+2} than in the interval

$$\{2^n + 1,\ 2^n + 2,\ \ldots,\ 2^{n+1}\}.$$

Here, as we have seen, there are $\binom{n}{2}$ good numbers, therefore

$$f(2^n + 2) = \binom{n}{2} + 1;$$

the equation $f(k) = m$ has one solution only if $m = \binom{n}{2} + 1$.

1994/4. *Determine all ordered pairs* (m, n) *of positive integers for which*

(1)
$$\frac{n^3 + 1}{mn - 1}$$

is an integer.

Solution. First we show that in spite of the apparent asymmetry, if a pair (m, n) makes (1) a whole number then so does the pair (n, m). Since m^3 is prime to $mn - 1$ the latter divides $n^3 + 1$, the numerator if and only if it also divides $m^3(n^3 + 1)$. On the other hand

$$m^3(n^3 + 1) - (m^3 + 1) = (mn)^3 - 1, \qquad \text{a multiple of} \quad (mn - 1)$$

therefore $m^3(n^3 + 1)$ and $m^3 + 1$ are congruent modulo $mn - 1$ and the claim follows. Hence we may assume that $m \geq n$ in (1).

Consider the case $m = n$ first. Then (1) becomes

$$\frac{n^3 + 1}{n^2 - 1} = n + \frac{1}{n - 1}$$

and this is a whole number only if $n = 2$; the pair $(2, 2)$ is hence a solution.

Let now $m > n$. If $n = 1$ then $\dfrac{2}{m - 1}$ is an integer only if $m = 2$ or $m = 3$; the pairs $(1, 2)$ and $(1, 3)$ are also solutions. If $n \geq 2$ then denoting the value of (1) by e we obtain

$$n^3 + 1 = emn - e, \qquad m = \frac{n^3 + 1 + e}{en}.$$

For m, to be an integer, $1 + e$ has to be divisible by n therefore $1 + e = kn$ for some positive integer k. Hence $e = kn - 1$ so (1) yields

$$kn - 1 = \frac{n^3 + 1}{mn - 1} < \frac{n^3 + 1}{n^2 - 1} = n + \frac{1}{n - 1}$$

that is

$$(k - 1)n < 1 + \frac{1}{n - 1}.$$

Since the *r.h.s.* is strictly less than 2, the only possibility is $k = 1$ and then

$$n^3 + 1 = (n - 1)(mn - 1).$$

Isolating m

$$m = \frac{n^2+1}{n-1} = n+1+\frac{2}{n-1},$$

which is an integer only if $n=2$ or $n=3$; the corresponding value of m is 5 in each case and thus the remaining solutions are the pairs $(2,5)$ and $(3,5)$.

Remembering our initial remark there are 9 solutions altogether, namely

$$(2,2), \quad (1,2), \quad (1,3), \quad (2,5), \quad (3,5),$$
$$(2,1), \quad (3,1), \quad (5,2), \quad (5,3).$$

1994/5. *Let S be the set of all real numbers greater than (-1). Find all functions f from S to S such that $f(x+f(y)+xf(y)) = y+f(x)+yf(x)$ for all x and y, $\dfrac{f(x)}{x}$ is strictly increasing on each of the intervals $-1 < x < 0$ and $0 < x$.*

Solution. Let $x = y \in S$. The functional equation hence becomes

(1) $$f(x+(1+x)f(x)) = x+(1+x)f(x).$$

Consider now the solutions of

(2) $$f(u) = u,$$

the so called fixed points of f. By (1) $u = x+(1+x)f(x)$ is a solution of (2) for every $x \in S$. If $u \neq 0$ then (2) can be written as

(3) $$\frac{f(u)}{u} = 1.$$

Since $\dfrac{f(u)}{u}$ is strictly monotone, it admits the value 1 at most once in both intervals $]-1,0[$ and $]0,+\infty[$.

Let $f(u) = u$ for some $u \in]-1,0[$. Substituting $x = u$ (1) becomes

$$f(u^2+2u) = u^2+2u.$$

Observe that the mapping $u \mapsto u^2+2u$ keeps the two parts of S fixed that is if $u \in]-1,0[$ then $u^2+2u \in]-1,0[$ and, similarly, if $u \in]0,+\infty[$ then $u^2+2u \in]0,\infty[$. Thus there are no solutions u_1, u_2 both in S such that $u_1 \in]-1,0[$, $u_2 \in]0,\infty[$ and $u_1^2+2u_1 = u_2$ and $u_2^2+2u_2 = u_1$.

Since there is at most one $u \in]-1,0[$ satisfying (2), if $u^2+2u \in]-1,0[$ then

$$u^2+2u = u.$$

But this quadratic has no solutions in $]-1,0[$, at all, and thus (2) cannot hold if $u \in]-1,0[$. A similar argument shows that (2) cannot hold in $]0,+\infty[$ either and thus $f(u) = u$ forces $u = 0$.

Thus (1) holds only if

$$x + (1+x)f(x) = 0 \qquad x \in S$$

and hence the only possibility is

(4)
$$f(x) = -\frac{x}{1+x}.$$

We have to check, of course, if this one is indeed a solution.

I. $f(x)$ is defined in S and

$$f(x) = -1 + \frac{1}{1+x}$$

shows that $x \in S$ implies $f(x) > -1$ that is $f(x) \in S$, indeed.

II.

$$\frac{f(x)}{x} = -\frac{1}{1+x}$$

is strictly increasing if $x > -1$.

III.

$$y + f(x) + yf(x) = y - \frac{x}{1+x} - \frac{xy}{1+x} = \frac{y-x}{1+x},$$

$$x + f(y) + xf(y) = \frac{x-y}{1+y},$$

$$f\left(\frac{x-y}{1+y}\right) = -\frac{\frac{x-y}{1+y}}{1+\frac{x-y}{1+y}} = -\frac{x-y}{1+x} = \frac{y-x}{1+x},$$

so $f(x) = -\dfrac{x}{1+x}$ satisfies the given functional equation and thus it is the one and only solution.

1994/6. *Show that there exists a set A of positive integers with the following property: for any infinite set of primes, there exist two positive integers m in A and n not in A, each of which is a product of k distinct elements of S for some $k \geq 2$.*

Solution. Define the set A as follows: it consists of those positive integers whose prime factorisation satisfies that

a) the index of any prime factor is 1, the elements of A are all *squarefree* numbers;

b) the number of prime factors is not less than the smallest prime divisor of the given number.

Let $S = \{p_1, p_2, \ldots\}$ where $p_1 < p_2 < p_3 < \ldots <$ is an arbitrary infinite set of primes.

Set k to be equal to p_1 $(p_1 \geq 2)$ and let

$$m = p_1 p_2 \cdots p_{p_1}, \qquad n = p_2 p_3 \cdots p_{p_1+1}.$$

Both these numbers are p_1-factor products and their factors are all from S. By the definition of A, the first number, m is in A. On the other hand, $n \in A$ is impossible otherwise n, having only p_1 prime factors, should possess p_2 of them, by definition, while $p_1 \geq p_2$. The set A as defined above thus satisfies the requirements.

1995.

1995/1. *Let A, B, C, D be four distinct points on a line, in that order. The circles with diameter AC and BD intersect at X and Y. Let P be a point on the line XY other than Z. The line CP intersects the circle with diameter AC at C and M, and the line BP intersects the circle with diameter BD at B and N. Prove that the lines AM, DN, XY are concurrent.*

First solution. The orthocentre O of $\triangle BPC$ is clearly incident to the straight line XY since the latter is perpendicular to BC. (*Figure 1995/1.1*). Point Z, on the other hand, is incident to the radical axis of the circles with diameters AC and BD therefore the respective powers are equal:

$$ZA \cdot ZC = ZB \cdot ZD, \qquad \text{that is} \qquad \frac{ZA}{ZB} = \frac{ZD}{ZC} = \lambda.$$

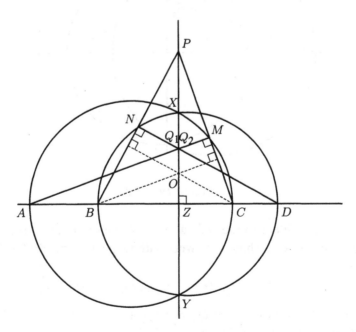

Figure 1995/1.1

AM and DN are parallel to BO and CO, respectively, since they are perpendicular to the same lines. Enlarge $\triangle BOC$ from Z by λ. This enlargement is mapping B to A, C to D and, at the same time, the straight lines BO and CO are mapped into AM and DN, respectively. Therefore O, the intersection of BO and CO is mapped into the intersection O' of the corresponding image lines. Hence the points Z, O and O' are collinear and thus AM, DN and XY are passing through the point O', they are concurrent, indeed.

Second solution. The circles with diameter AC and BD are denoted by c_1 and c_2, respectively (*Figures 1995/1.2* and *3.*) intersections of AM and DN with XY be Q_1 and Q_2. Thus we have to show that these two points, Q_1 and Q_2 are identical. By *Thales'* theorem $\angle AMC = \angle BND = 90°$. The angles at M and Z subtended by the chord $Q_1 C$ are both right angles therefore the points Z, C, M, Q_1 are lying on a circle c_3. A similar argument yields that the four points Z, B, N, Q_2 are incident to a circle c_4.

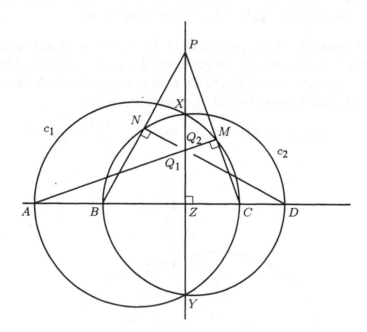

Figure 1995/1.2

P is hence incident to the radical axes of the circle pairs (c_1, c_2), (c_1, c_3) and $(c_2 \, c_4)$ respectively. Thus its powers with respect to the circles c_3 and c_4 are also equal:

$$PQ_1 \cdot PZ = PQ_2 \cdot PZ, \qquad \text{yielding} \qquad PQ_1 = PQ_2.$$

We are to finish now; since Q_1 and Q_2 are not separated by P on the line XY, the two points, Q_1 and Q_2 are identical, indeed.

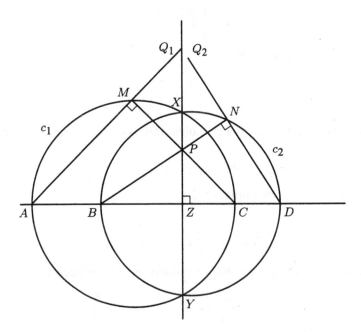

Figure 1995/1.3

Remark. Two different configurations due to the position of P are shown in the respective figures; the apparent difference, however, has no effect on the proof.

1995/2. *Let a, b, c be positive real numbers with $abc = 1$. Prove that*
$$1/a^3(b+c) + 1/b^3(c+a) + 1/c^3(a+b) \geq 3/2.$$

First solution. Denote the sum on the *l.h.s.* by S and apply *Cauchy's inequality* ([22]) for the two awkward triples
$$\left(\sqrt{1/a^3(b+c)},\ \sqrt{1/b^3(c+a)},\ \sqrt{1/c^3(a+b)} \right)$$
and
$$\left(\sqrt{b+c/bc},\ \sqrt{c+a/ca},\ \sqrt{a+b/ab} \right).$$

We thus obtain
$$\left(\sqrt{\frac{1}{a^3bc}} + \sqrt{\frac{1}{ab^3c}} + \sqrt{\frac{1}{abc^3}} \right)^2 \leq S \left(\frac{b+c}{bc} + \frac{c+a}{ca} + \frac{a+b}{ab} \right),$$
$$\left(\frac{1}{a} + \frac{1}{b} + \frac{1}{c} \right)^2 \leq 2S \left(\frac{1}{a} + \frac{1}{b} + \frac{1}{c} \right);$$

that is

$$S \geq \frac{1}{2}\left(\frac{1}{a} + \frac{1}{b} + \frac{1}{c}\right).$$

Going on with the A.M.–G.M. inequality:

$$S \geq \frac{3}{2}\frac{\frac{1}{a}+\frac{1}{b}+\frac{1}{c}}{3} \geq \frac{3}{2}\sqrt[3]{\frac{1}{abc}} = \frac{3}{2}.$$

·By the last estimation equality holds only if $a = b = c$ and then it clearly does.

Second solution. With the notations $x = \frac{1}{a}$, $y = \frac{1}{b}$, $z = \frac{1}{c}$ we have $xyz = 1$. Now the first term, for example, of the *l.h.s.* is

$$\frac{1}{a^3(b+c)} = \frac{x^3}{\frac{1}{y}+\frac{1}{z}} = \frac{x^3yz}{y+z} = \frac{x^2}{y+z}.$$

Rewriting the remaining two terms similarly the claim becomes

$$(1) \qquad S = \frac{x^2}{y+z} + \frac{y^2}{z+x} + \frac{z^2}{x+y} \geq \frac{3}{2};$$

we shall proceed by proving (1).

We need the extended A.M.–H.M. inequality ([40]) for weighted means; the terms are now $\dfrac{x}{y+z}$, $\dfrac{y}{z+x}$, $\dfrac{z}{x+y}$ with the respective weights x, y, z:

$$\frac{S}{x+y+z} = \frac{x\frac{x}{y+z}+y\frac{y}{z+x}+z\frac{z}{x+y}}{x+y+z} \geq \frac{x+y+z}{x\frac{y+z}{x}+y\frac{z+x}{y}+z\frac{x+y}{z}},$$

that is

$$S \geq \frac{(x+y+z)^2}{2(x+y+z)} = \frac{x+y+z}{2}.$$

The A.M.–G.M. inequality settles the problem again; since the product of the new variables is 1, we get

$$S \geq \frac{3}{2}\frac{x+y+z}{3} \geq \frac{3}{2}\sqrt[3]{xyz} = \frac{3}{2}.$$

Equality holds if and only if $x = y = z$ that is $a = b = c$.

Third solution. There are various generalizations of the problem; we show one of these proving hence the claim again.

Additionally to the given conditions let $\beta \geq 2$; we are going to prove the inequality

$$\frac{1}{a^\beta(b+c)} + \frac{1}{b^\beta(c+a)} + \frac{1}{c^\beta(a+b)} \geq \frac{3}{2}.$$

Introducing, as in the previous solution, the variables $x = \dfrac{1}{a}$, $y = \dfrac{1}{b}$, $z = \dfrac{1}{c}$ the claim becomes

$$\frac{x^{\beta-1}}{y+z} + \frac{y^{\beta-1}}{z+x} + \frac{z^{\beta-1}}{x+y} \geq \frac{3}{2}.$$

To make it simpler let $\alpha = \beta - 1$ $(\alpha \geq 1)$; what we have to prove now is

$$S_\alpha = \frac{x^\alpha}{y+z} + \frac{y^\alpha}{z+x} + \frac{z^\alpha}{x+y} \geq \frac{3}{2} \qquad (\alpha \geq 1).$$

First we check the case $\alpha = 1$ (we don't need $xyz = 1$ now).

$$S_1 = \frac{x}{y+z} + \frac{y}{z+x} + \frac{z}{x+y} = \frac{x+y+z}{y+z} + \frac{x+y+z}{z+x} + \frac{x+y+z}{x+y} - 3 =$$

$$= 3(x+y+z) \cdot \frac{\dfrac{1}{y+z} + \dfrac{1}{z+x} + \dfrac{1}{x+y}}{3} - 3.$$

The A.M.–H.M. inequality now yields

$$S_1 \geq 3(x+y+z)\frac{3}{y+z+z+x+x+y} - 3 = \frac{9(x+y+z)}{2(x+y+z)} - 3 = \frac{3}{2}.$$

Assume now that $x \geq y \geq z$; then clearly $x^{\alpha-1} \geq y^{\alpha-1} \geq z^{\alpha-1}$ and the inequalities

$$\frac{x}{y+z} \geq \frac{y}{z+x} \geq \frac{z}{x+y}$$

also hold. Indeed, $x \geq y$, $y + z \leq z + x$ and $\dfrac{1}{y+z} \geq \dfrac{1}{z+x}$, therefore

$$\frac{x}{y+z} \geq \frac{y}{x+z}.$$

We use now *Chebyshev's inequality*. It states that if

$$a_1, a_2, \ldots, a_n \qquad \text{and} \qquad b_1, b_2, \ldots, b_n$$

are two, similarly ordered sequences of real numbers then

$$\frac{a_1 b_1 + a_2 b_2 + \ldots + a_n b_n}{n} \geq \frac{a_1 + a_2 + \ldots + a_n}{n} \cdot \frac{b_1 + b_2 + \ldots + b_n}{n}.$$

(If the sequences are arranged in opposite order then the inequality holds the other way round; we have equality if and only if $a_1 = a_2 = \ldots = a_n$ or $b_1 = b_2 = \ldots = b_n$.)

Applying Chebyshev's inequality for the triples

$$x^{\alpha-1}, \ y^{\alpha-1}, \ z^{\alpha-1} \qquad \text{and} \qquad \frac{x}{y+z}, \ \frac{y}{z+x}, \ \frac{z}{x+y}$$

yields

$$S_\alpha \geq S_1 \frac{x^{\alpha-1} + y^{\alpha-1} + z^{\alpha-1}}{3} \geq \frac{3}{2} \frac{x^{\alpha-1} + y^{\alpha-1} + z^{\alpha-1}}{3}.$$

The end is A.M.–G.M. again. Since $xyz = 1$:

$$S_\alpha \geq \frac{3}{2}\sqrt[3]{(xyz)^{\alpha-1}} = \frac{3}{2},$$

and equality holds if and only if $x = y = z$ that is $a = b = c$.

1995/3. *Determine all integers $n > 3$ for which there exist n points A_1, A_2, ..., A_n in the plane, no three collinear and real numbers r_1, r_2, ..., r_n such that for any distinct i, j, k, the area of the triangle $A_i A_j A_k$ is $r_i + r_j + r_k$.*

Solution. We prove that the requirements can be satisfied only if $n = 4$.

If this is the case then consider a unit square with vertices A_1, A_2, A_3, A_4 and let $r_1 = r_2 = r_3 = r_4 = \frac{1}{6}$. Now any triple of the given points forms a triangle of area $3 \cdot \frac{1}{6} = \frac{1}{2}$. (Instead of a square any parallelogram of unit area will do.)

Next we prove that if there are five points given then there can be no equal ones among the numbers r_i. Assume, to the contrary, that, for example $r_4 = r_5$. Then $[A_1 A_2 A_4] = r_1 + r_2 + r_4$ and $[A_1 A_2 A_5] = r_1 + r_2 + r_4$; the two areas are equal, therefore $A_1 A_2$ and $A_4 A_5$ are parallel. A similar argument for the triangles $A_2 A_3 A_4$ and $A_2 A_3 A_5$ yields that under the assumption $r_4 = r_5$ the lines $A_2 A_3$ and $A_4 A_5$ are also parallel. Now this is already a contradiction since it means that the points A_1, A_2, A_3 are collinear. (*Figure 1995/3.1*).

Assume now that there are five points on the plane with the required property. If their convex hull is a pentagon (*Figure 1995/3.2*) then, clearly,

$$[A_1 A_2 A_3 A_4] = [A_1 A_2 A_3] + [A_1 A_4 A_3] = [A_2 A_3 A_4] + [A_4 A_1 A_2] \quad \text{that is}$$

$$r_1 + r_2 + r_3 + r_1 + r_4 + r_3 = r_2 + r_3 + r_4 + r_4 + r_1 + r_2, \quad \text{and hence}$$

$$r_1 + r_3 = r_2 + r_4.$$

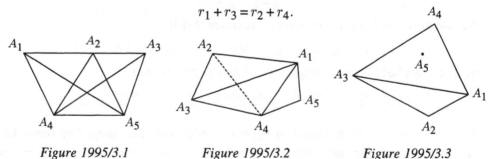

Figure 1995/3.1   Figure 1995/3.2   Figure 1995/3.3

The same argument when applied to the convex quadrilateral $A_1 A_2 A_3 A_5$, yields

$$r_1 + r_3 = r_2 + r_5$$

and thus $r_4 = r_5$, which we have already seen to be impossible.

If the convex hull of the points is a quadrilateral, $A_1 A_2 A_3 A_4$ for example, then A_5 is an interior point of the convex hull (*Figure 1995/3.3*). The previous argument can be repeated, without essential modification, to the quadrilaterals $A_1A_2A_3A_4$ and $A_1A_2A_3A_5$ yielding once more the impossible $r_4 = r_5$.

Finally, if the convex hull is the triangle $A_1A_2A_3$ then clearly

$$[A_1A_2A_4] + [A_2A_3A_4] + [A_3A_1A_4] = [A_1A_2A_5] + [A_2A_3A_5] + [A_3A_1A_5]$$

that is

$$r_1 + r_2 + r_5 + r_2 + r_3 + r_5 + r_3 + r_1 + r_5 = r_1 + r_2 + r_4 + r_2 + r_3 + r_4 + r_3 + r_1 + r_4,$$

and hence

$$r_5 = r_4,$$

again. There are no five points (and more, of course) satisfying the conditions.

1995/4. *Find the maximum value of x_0 for which there exists a sequence $x_0, x_1, \ldots, x_{1995}$ of positive reals with $x_0 = x_{1995}$ such that for $i = 1, 2, \ldots, 1995$*

(i) $$x_0 = x_{1995};$$

(ii) $$x_{i-1} + \frac{2}{x_{i-1}} = 2x_i + \frac{1}{x_i}.$$

Solution. With a bit of algebra, relation (ii) can be written in more tractable form as a quadratic in x_i:

$$x_i^2 - \left(\frac{x_{i-1}}{2} + \frac{1}{x_{i-1}} \right) x_i + \frac{1}{2} = 0,$$

and hence

(1) $x_i = \dfrac{x_{i-1}}{2}$ or   (2) $x_i = \dfrac{1}{x_{i-1}}.$

This shows that starting from x_0 each term of the sequence can be obtained from the previous one by either halving it or as its reciprocal. Therefore, each term of the sequence is either of the form

(3) $$x_i = 2^k x_0, \qquad \text{or} \qquad x_i = \frac{2^k}{x_0}$$

for some integer k.

Hence, by (i), the 1995th term satisfies either

(4) $x_{1995} = 2^k x_0 = x_0$,   or   (5) $x_{1995} = \dfrac{2^k}{x_0} = x_0$, and thus $x_0^2 = 2^k$.

Arriving from x_0 to x_{1995} the iteration steps (1) and (2) have been performed 1995 that is an odd number of times altogether. (4) can be reached only if (2) has been applied even times that leaves odd steps of type (1). Observe that applying (1) changes the index of 2 by 1 and, at the beginning we have $x_0 = 2^0 x_0$. Therefore if $x_{1995} = 2^k x_0$ then k has to be odd. Since $x_0 > 0$, x_{1995} now cannot be equal to x_0.

To arrive to (5) step (2) has to be applied at least once and thus (1) can be used at most 1994 times. Hence 2^k, the power factor in (5) cannot exceed 2^{1994} and thus the maximum value of x_0^2 is at most 2^{1994}. Hence

$$x_0 \leq \sqrt{2^{1994}} = 2^{997}.$$

On the other hand $x_0 = 2^{997}$ is already possible. For this the first 1994 iteration steps are all of the first kind and the last step is (2); the sequence produced from the maximal value of x_0 is hence

$$x_0 = 2^{997}, \quad x_1 = 2^{996}, \quad \ldots, \quad x_{997} = 2^0 = 1, \quad x_{998} = 2^{-1}, \quad \ldots, \quad x_{1994} = 2^{-997},$$
$$x_{1995} = 2^{997}.$$

Remark. Calculating the first few terms we can verify that

$$x_i = 2^{k_i} x_0^{\varepsilon_i},$$

where $|k_i| \leq i$ and $\varepsilon_i = (-1)^{i+k_i}$; this can be proved by mathematical induction but it also follows from the argument of the solution.

1995/5. *Let $ABCDEF$ be a convex hexagon with $AB = BC = CD$ and $DE = EF = FA$ such that $\angle BCD = \angle EFA = 60°$. Suppose that G and H are points in the interior of the hexagon such that $\angle AGB = \angle DHE = 120°$. Prove that*

(1) $$AG + GB + GH + DH + HE \geq CF.$$

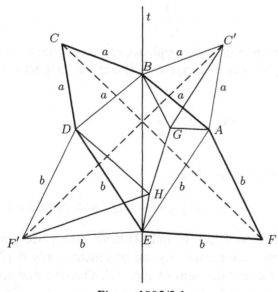

Figure 1995/5.1

Solution. Let $AB = a$ and $AE = b$. By the conditions $\triangle BCD$ and $\triangle EFA$ are equilateral triangles of sides a and b respectively. Draw equilateral triangles $AC'B$ and $DF'E$ externally upon the sides AB and DE respectively (*Figure 1995/5.1*). Since $ABDE$ is a kite of sides a and b the whole figure is symmetrical to the axis $BE = t$ of the kite. Hence $CF = C'F'$.

G and H are incident to the circumcircles of the equilateral triangles $AC'B$ and $DF'E$ respectively and thus, by ([43])

$$GC' = AG + GB, \qquad HF' = DH + HE.$$

Hence the sum on the *l.h.s.* of (1) is equal to the length of the broken line $C'GHF'$. Since the latter is clearly at least $C'F'$, the *l.h.s.* of (1) is at least $C'F' = CF$ and the proof is hence finished.

It is clear that equality holds in (1) if and only if H and G are incident to $C'F'$.

1995/6. *Let p be an odd prime number. How many p-element subsets A of $\{1, 2, \ldots, 2p\}$ are there, the sum of whose elements is divisible by p?*

Solution. Split $H = \{1, 2, \ldots, 2p\}$ into two p-element subsets as follows

$$A = \{1, 2, \ldots, p\}, \qquad B = \{p+1, p+2, \ldots, 2p\}.$$

Any p-element subset C of H different from both A and B contains elements from both A and B; consider its intersection with A:

$$a_1, a_2, \ldots, a_n \qquad (1 \leq n \leq p-1).$$

Add now 1 to each element of $A \cap C$ and replace every number thus modified in C by its remainder when divided by p; if this remainder happens to be zero then write p instead.

Denote by C_1 the set obtained this way from C. Similarly: adding 2, 3 $\ldots$ $\ldots, p$ to the elements a_i inside C respectively and reducing them $b\!\!\mod p$ in the above sense yields the p-element subsets $C_2, C_3, \ldots, C_p$; the set C_p is clearly identical to C itself. These subsets form a closed class of the p-element subsets of H, any one of them generates the class via the above procedure and subsets outside of this class generate different subsets.

For an arbitrary p-element subset C of H denote the sum of its elements by $s(C)$. $s(A)$, for example, is equal to $\dfrac{p(p+1)}{2}$, $s(B) = s(A) + p^2$. Both $s(A)$ and $s(B)$ are divisible by p because it is odd. Since the respective sums of the elements in C_i and C_{i+1} differ by n (mod p, of course) we have the following congruence:

$$s(C_i) - s(C_j) \equiv (i - j)n \pmod{p}.$$

$|i - j| < p$ and $n < p$ imply that $(i - j)n$ is not a multiple of p; hence the numbers $s(C_i)$ belong to different residue classes mod p. Since there are p members of the C-class there is exactly one C_i such that $s(C_i) \equiv 0 \pmod{p}$, the sum of whose elements is divisible by p.

Apart from A and B there are $\dbinom{2p}{p} - 2$ p-element subsets of H. These subsets, as we have seen, can be arranged into $\dfrac{1}{p}\left(\dbinom{2p}{p} - 2\right)$ classes and there is a single subset in each class the sum of whose elements is divisible by p. Therefore the number of p-element subsets (A and B included) of H where the sum of the elements is divisible by p is

$$\frac{\binom{2p}{p} - 2}{p} + 2.$$

Note. An interesting arithmetical byproduct of the result is that $\binom{2p}{p} - 2$ a multiple of p whenever p is an odd prime; indeed, by its meaning, the ratio in the result is an integer.

1996.

1996/1. *We are given a positive integer r and a rectangular board divided into 20×12 unit squares. The following moves are permitted on the board: one can move from one square to another only if the distance between the centres of the two squares is $\sqrt{r}$. The task is to find a sequence of moves leading between two adjacent corners of the board which lie on the long side.*

(a) *Show that the task cannot be done if r is divisible by 2 or 3.*

(b) *Prove that the task is possible for $r = 73$.*

(c) *Can the task be done for $r = 97$?*

Solution. Label its corners and place the board $ABCD$ in a Cartesian system with the centre of the A-square at the origin, the centre of the B-square at (19, 0), finally, let the centre of the D-square be (11, 0). The moves can be resolved into the sum of horizontal and vertical components and thus we may assume that our token is moving parallel to the axes.

(a) Let r be even first and consider a vector $\mathbf{v}(x, y)$ for a legal move; then, by condition, $x^2 + y^2 = r$. Its value is now even and thus $x \equiv y \pmod 2$. Colour the squares as on a usual chessboard with the field of A black. If x and y are both even then any move, horizontal or vertical, takes us to a field of the same colour; if both components are odd then the colour of the field flips step by step and thus, eventually, the destination and the start have the same colour again. Since B, the target square is white, no way to reach it from the black A.

1 Let $r = x^2 + y^2$ be now divisible by 3. Since a square mod 3 is equal to 0 or 1, the sum of two squares is divisible by 3 if and only if both x and y are of the kind. Draw a 3×3 grid on the infinite board and colour the 9 fields of each square of the grid by one of three colours according to the pattern:

2	3	1
3	1	2
1	2	3

A

The length of any move, horizontal or vertical, is now divisible by 3 and thus we are travelling along squares of the same colour, all the time. On the other hand, if A gets colour 1 then B has colour 2, therefore it cannot be reached from A.

(b) There is just one way to split $r = 73$ into the sum of two squares: $73 = = 8^2 + 3^2$. Thus every legal move is by one of the following vectors:

$$(3, 8), \ (8, 3), \ (-8, 3), \ (3, -8), \ (-3, 8), \ (8, -3), \ (-3, -8), \ (-8, -3)$$

The task is to produce $(19, 0)$ as the sum of vectors from this list; the token, during the trip, is not allowed to exit from the board, of course. One can find such a trip with trial and error (*Figure 1996/1.1*); the list below displays the stepwise positions of the moving token, the corresponding steps are indicated above the arrows, respectively.

$$(0, 0) \xrightarrow{(3,8)} (3, 8) \xrightarrow{(8,-3)} (11, 5) \xrightarrow{(8,-3)} (19, 2) \xrightarrow{(-3,8)} (16, 10) \xrightarrow{(-8,-3)} (8, 7) \rightarrow$$
$$\xrightarrow{(-8,-3)} (0, 4) \xrightarrow{(8,-3)} (8, 1) \xrightarrow{(3,8)} (11, 9) \xrightarrow{(-8,-3)} (3, 6) \xrightarrow{(8,-3)} (11, 3) \xrightarrow{(8,-3)} (19, 0).$$

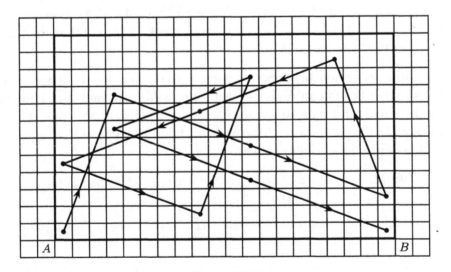

Figure 96/1.1

(c) 97 can be written as the sum of two squares only as $9^2 + 4^2$ and thus the length of any move parallel to the axes is 4 or 9. Colour the fields of the board once more, according to the pattern on *Figure 1996/1.2*; note again that the A-square is black and the target B is white.

The argument should be familiar by now: starting from a black field, a horizontal step by 4 takes us to a black field again and so does a vertical step of 9; a horizontal 9 is white but a vertical 4 is black again. The story is the same with a vertical 4 start: ending up on a white square, a horizontal 9 returns to a black field; finally, moving vertically by 9 and horizontally by 4 we are not

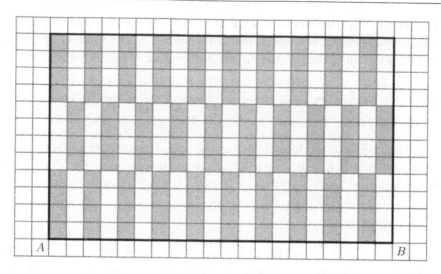

Figure 96/1.2

leaving the safe black area at all. Coming to the end: if we start from a black field then every legal move will take us to black fields and thus the white B is inaccessible from the black A.

Note. A reasonably simple argument shows that the 11 long sequence of moves presented in part (b) is the shortest possible trip from A to B; the moves, by the way, can be found by an algorithmic approach, although this was not required in the competition. Most students have quicly found one of the several solutions with bare hands, anyway.

1996/2. *Let P be a point inside the triangle ABC such that*

(1) $$\angle APB - \angle ACB = \angle APC - \angle ABC.$$

Let D, E be the incentres of triangles APB, APC respectively. Show that AP, BD and CE meet at a point.

First solution. Since D and E are incentres, BD and CE are bisecting the angles $\angle ABP$ and $\angle ACP$. By the angle bisector theorem these two lines meet on the segment AP if and only if they divide it into parts of equal ratios, respectively. This common ratio is hence equal to the ratio of the sides enclosing AP (*Figure 1996/2.1*). This happens if

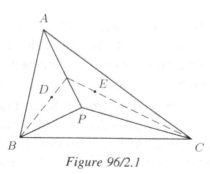

Figure 96/2.1

(2) $$\frac{AB}{PB} = \frac{AC}{PC}, \quad \text{that is} \quad \frac{AB}{AC} = \frac{PB}{PC}.$$

We solve the problem by proving the latter equality.

The lines AP, BP and CP meet the circumcircle of $\triangle ABC$ for the second time at X, Y and Z, respectively (*Figure 1996/2.2*). The very same lines are dividing the angles α, β, γ of $\triangle ABC$ into α_1, α_2; β_1, β_2; γ_1, γ_2, respectively, as it is shown in the figure. By the theorem of inscribed angles we can write down the angles of $\triangle XYZ$:

$$\angle X = \beta_1 + \gamma_2, \qquad \angle Y = \alpha_2 + \gamma_1, \qquad \angle Z = \alpha_1 + \beta_2.$$

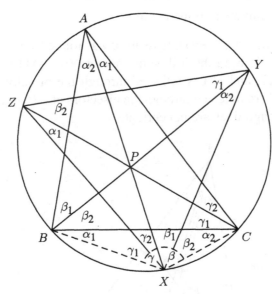

Figure 96/2.2

$\angle APB$, on the other hand, is the sum of the respective exterior angles of $\triangle ACP$ and $\triangle BCP$; thus

$$\angle APB - \angle ACB = \alpha_1 + \gamma_2 + \beta_2 + \gamma_1 - \gamma_1 - \gamma_2 = \alpha_1 + \beta_2.$$

Similarly,

$$\angle APC - \angle ABC = \alpha_2 + \beta_1 + \beta_2 + \gamma_1 - \beta_1 - \beta_2 = \alpha_2 + \gamma_1,$$

and hence, by (1)

(3) $$\alpha_1 + \beta_2 = \alpha_2 + \gamma_1.$$

Thus, in $\triangle XYZ$ we have $\angle Z = \angle Y$, it is isosceles,

(4) $$XY = ZX.$$

Observe now that the pairs $\triangle ABP$, $\triangle YXP$ and $\triangle ACP$, $\triangle ZXP$ are respectively similar because the corresponding angles are equal in each pair. By these similarities

$$\frac{AB}{PB} = \frac{XY}{XP}, \qquad \text{abd} \qquad \frac{AC}{PC} = \frac{ZX}{XP},$$

which, by (4), yields $\dfrac{AB}{PB} = \dfrac{AC}{PC}$, the claim.

Second solution. We use some pieces from the previous solution. Reviewing *Figure 1996/2.2* we can check that in $\triangle PBX$ and $\triangle PCX$ we have $\angle PXB = \gamma$ and $\angle PXC = \beta$ and also $\angle PBX = \alpha_1 + \beta_2$ and $\angle PCX = \alpha_2 + \gamma_1$. This, by (3), yields $\angle PBX = \angle PCX$. Apply the sine rule in $\triangle PBX$, $\triangle PCX$, finally in $\triangle ABC$:

$$\frac{PB}{PC} = \frac{PB}{PX} \cdot \frac{PX}{PC} = \frac{\sin\gamma}{\sin PBX\angle} \cdot \frac{\sin PCX\angle}{\sin\beta} = \frac{\sin\gamma}{\sin\beta} = \frac{AB}{AC}$$

which, by (2), is equivalent to the claim.

Third solution. An interesting relation was found by *Peter Frenkel*, a Hungarian contestant. He used the following theorem (see [41]): if P is an interior point of the triangle and PA is reflected through the bisector of $\angle A$ and, similarly, PB and PC are reflected through the bisectors of $\angle B$ and $\angle C$, respectively, then the mirror image lines are concurrent.

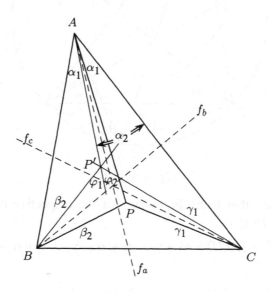

Figure 96/2.3

The straight lines PA, PB, PC are dividing the respective angle of $\triangle ABC$ as it is shown in *Figure 1996/2.2*. Reflect now the lines PA, PB, PC as in the above theorem, through the bisectors f_a, f_b and f_c, respectively. The image lines, by the assertion, meet at some point P' (*Figure 1996/2.3*). Because of reflection clearly $\angle P'BA = \beta_2$ and $\angle P'CA = \gamma_1$. The straight line AP' divides $\angle BP'C$ into the parts φ_1 and φ_2; hence, by the exterior angle relation

$$\varphi_1 = \alpha_1 + \beta_2, \qquad \varphi_2 = \alpha_2 + \gamma_1,$$

and this, by (3), yields $\varphi_1 = \varphi_2$ that is AP' is bisecting $\angle BP'C$.

Denote the incentre of $\triangle BP'C$ by K. Apply the theorem again: the mirror images of the lines KA, KB, KC through the respective angle bisectors of

$\triangle ABC$ are thus concurrent. The lines KA and $P'A$ are identical and their common mirror image through f_a is PA. Since reflection is mapping angle bisectors into angle bisectors, the mirror image of KB is DB and that of KC is EC; the claim hence has been proved.

Remark. According to relation (2) the ratio of the respective distances of the point P of the problem, from the vertices B and C is equal to c/b, The locus of such points is the so called *Apollonius' circle*; its diameter is the segment cut from BC by the bisectors of the interior and the exterior angle at A.

1996/3. *Let S be the set of non-negative integers. Find all functions f from S to itself such that*

(1) $$f(m + f(n)) = f(f(m)) + f(n)$$

for all m, n.

Solution. If $m = n = 0$ then (1) implies
$$f(f(0)) = f(f(0)) + f(0)$$
that is $f(0) = 0$. Set now $m = 0$ in (1):
$$f(f(n)) = f(f(0)) + f(n) = f(n).$$
Hence f keeps the elements of its range fixed,

(2) $$f(f(n)) = f(n).$$

Therefore, (1) can be rewritten as:

(3) $$f(m + f(n)) = f(m) + f(n);$$

in what follows we shall be using this property of f.

The constant zero function is clearly a solution so we assume that in what follows f is not all zero. Denote the smallest positive fixed point of f by a. This value, by (2) and our previous assumption, is well defined. We prove, by induction on k, that ka is also a fixed point for every positive integer k. For $k = 1$ this is the definition of a. Let $k > 1$ and assume that $f((k-1)a) = (k-1)a$.

Substituting $m = a$ and $n = (k-1)a$ in (3) yields
$$f(ka) = f(a + f(k-1)a) = f(a) + f((k-1)a) = a + (k-1)a = ka$$
so ka is also kept fixed by f.

Using the minimum property of a in the usual way we now show that every fixed point of f is equal to some positive multiple of a. Indeed, if $b = aq + r$ ($q > 0$, $0 \leq r < a$ are integers) is an arbitrary fixed point then, by the previous result, qa likewise and thus, by (3)
$$aq + r = b = f(b) = f(aq + r) = f(r + f(aq)) = f(r) + f(aq) = f(r) + aq,$$
yielding
$$f(r) = r.$$

Thus r is also a fixed point and being smaller than a, the smallest positive fixed point, it is forced to be zero, $b = aq$, indeed. Making the two ends meet, since the elements of the range are all kept fixed, $f(n)$ is equal to ka with some k for every n.

If $0 \le i < a$ then let $f(i) = n_i a$ (n_i is a positive integer, $n_0 = 0$). If $n = ka + i$ is an arbitrary integer ($0 \le i < a$) then, by (3)

$$f(n) = f(ka + i) = f(i + f(ka)) = f(i) + f(ka) = n_i a + ka,$$

and thus

(4) $$f(n) = f(ka + i) = (n_i + k)a \qquad (0 \le i < a).$$

Now we are ready to characterize the solutions of (1). Set an arbitrary positive integer a as the smallest fixed point of f and choose the $a - 1$ values $f(1) = n_1$, $f(2) = n_2, \ldots, f(a-1) = n_{a-1}$ from S arbitrarily, apart from the restriction $i \ne$ $\ne n_i$. Set, finally, $f(0) = 0$. By (4) these functions can uniquely be extended to S.

To finish properly we are left to show that these functions do satisfy (1). For this let

$$m = k_1 a + i, \qquad n = k_2 a + j \qquad (0 \le i, j < a).$$

Now clearly

$$f(m + f(n)) = f(k_1 a + i + n_j a + k_2 a) = f((k_1 + k_2 + n_j)a + i) =$$
$$= (k_1 + k_2 + n_i + n_j)a,$$
$$f(f(m)) + f(n) = f((k_1 + n_i)a) + f(k_2 a + j) = (k_1 + n_i)a + (k_2 + n_j)a =$$
$$= (k_1 + k_2 + n_i + n_j)a,$$

and thus (1) does hold, indeed.

Remark. Since for $n = ka + i$ clearly $k = \left[\dfrac{n}{a}\right]$, (4) can also be written as

$$f(n) = \left(\left[\frac{n}{a}\right] + n_i\right)a.$$

1996/4. *The positive integers a and b are such that $15a + 16b$ and $16a - 15b$ are both squares of positive integers. What is the least possible value that can be taken by the smaller of these two squares?*

Solution. If $15a + 16b = r^2$ and $16a - 15b = s^2$ for some whole numbers r and s then

$$r^4 + s^4 = 225(a^2 + b^2) + 256(a^2 + b^2) + 2 \cdot 15 \cdot 16ab - 2 \cdot 15 \cdot 16ab =$$
$$= 481(a^2 + b^2) = 13 \cdot 37(a^2 + b^2).$$

$(13, 37) = 1$, therefore $r^4 + s^4$ is divisible by both 13 and 37.

Simple case checking yields the possible remainders of an arbitrary 4th power when divided by 13; they are

$$0, \quad 1, \quad 3, \quad 9.$$

The sum of two 4th powers is hence divisible by 13 if and only if both terms are of the kind.

The story is similar if the remainders of 4th powers when divided by 37 are checked; they are

$$0, \quad 1, \quad 7, \quad 9, \quad 10, \quad 12, \quad 16, \quad 26, \quad 33, \quad 34.$$

Being a multiple of 13 and 37, $r^4 + s^4$ thus forces both r and s to be a multiple of these numbers and hence that of 481, their product. Thus

$$r \geq 481, \qquad s \geq 481.$$

Observe, on the other hand, that 481, the smallest possible value of r and s can, in fact, be realized because the system

$$15a + 16b = 481^2,$$

$$16a - 15b = 481^2$$

can be solved in the set of positive integers. Indeed

$$a = 481 \cdot 31 = 14\ 911, \qquad b = 481$$

and thus the least positive value of the smaller one of the two squares is

$$481^2 = 231\ 361.$$

Remark. Fermat's small theorem ([32]) provides a uniform way to prove that $x^4 + y^4$ is divisible by 13 and 37 respectively only if both x and y are divisible by these numbers.

It is obvious that if $x^4 + y^4$ is divisible by 13 and any one of x and y also has this property then so does the other one. Assume hence that none of the two numbers are divisible by 13 but the sum of their 4th powers is.

$x^4 + y^4 \equiv 0$ (mod 13) implies $x^4 \equiv -y^4$; then also $x^{12} \equiv -y^{12}$. By the small Fermat theorem, however,

$$x^{12} \equiv 1 \quad \text{and} \quad y^{12} \equiv 1 \quad (\text{mod } 13),$$

and thus $1 \equiv -1$ or $2 \equiv 0$, a contradiction.

The point is that (13-1)/4 is an odd number and thus the signs of the opposite terms in the congruence relation remain opposite after the index has reached 12. The same proof works whenever p is a prime of the form $4k+1$ with k odd, like 37.

1996/5. *Let $ABCDEF$ be a convex hexagon of perimeter p such that AB is parallel to DE, BC is parallel to EF and CD is parallel to FA. Let R_A, R_C, and R_E denote the circumradii of triangles FAB, BCD, DEF respectively.*

Prove that

(1) $$R_A + R_C + R_E \geq \frac{p}{2}.$$

Solution. Since the opposite sides of the hexagon are given to be parallel, $\angle A = \angle D = \alpha$, $\angle B = \angle E = \beta$ and $\angle C = \angle F = \gamma$. Construct a rectangle $XYZU$

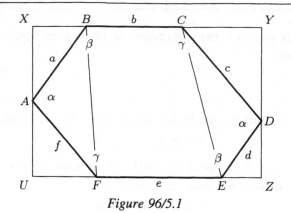

Figure 96/5.1

about the hexagon by drawing perpendiculars to the lines BC and EF from the vertices A and D, respectively (*Figure 1996/5.1*).

With $AB = a$, $BC = b$, $CD = c$, $DE = d$, $EF = e$, $FA = f$ clearly

$$XA = a \sin B, \quad AU = f \sin C, \quad YD = c \sin C, \quad DZ = d \sin A.$$

Observe that $BF \geq XU = YZ$ and thus $2BF \geq XU + YZ$; therefore

$$2BF \geq XA + AU + YD + DZ = (a+d) \sin B + (c+f) \sin C.$$

The circumradius of $\triangle FAB$ is

$$R_A = \frac{BF}{2 \sin A} = \frac{2BF}{4 \sin A} \geq \frac{1}{4} \left((a+d) \frac{\sin B}{\sin A} + (c+f) \frac{\sin C}{\sin A} \right).$$

Similarly

$$R_C \geq \frac{1}{4} \left((c+f) \frac{\sin A}{\sin C} + (b+e) \frac{\sin B}{\sin C} \right) \quad and$$

$$R_E \geq \frac{1}{4} \left((b+e) \frac{\sin C}{\sin B} + (a+d) \frac{\sin \alpha}{\sin B} \right).$$

Adding these inequalities

$$4(R_A + R_C + R_E) \geq$$

$$\geq (a+d) \left(\frac{\sin A}{\sin B} + \frac{\sin B}{\sin A} \right) + (c+f) \left(\frac{\sin A}{\sin \gamma} + \frac{\sin \gamma}{\sin A} \right) + (b+e) \left(\frac{\sin C}{\sin B} + \frac{\sin B}{\sin \gamma} \right).$$

Since the sum of a positive number and its reciprocal is at least 2,

$$4(R_A + R_C + R_E) \geq 2(a+b+c+d+e+f) = 2p,$$

$$R_A + R_C + R_E \geq \frac{p}{2},$$

indeed. Equality holds if and only if the hexagon is regular.

Remark. The problem is closely related to the *Erdős–Mordell inequality* ([33]) and its extension.

Construct the parallelograms $FABP$, $BCDQ$, $DEFS$ inside the hexagon and draw perpendiculars from the vertices F, B, D to the lines PF, QB, SD,

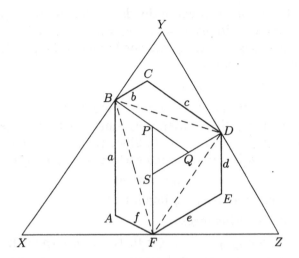

Figure 96/5.2.

respectively. These perpendiculars enclose a triangle XYZ. X is the intersection of the lines perpendicular to PF and QB and so on, see *Figure 1996/5.2.*

$XFPB$ is cyclic because a right angle is subtended by PX at both F and B. The circumradius of this quadrilateral is equal to that of $\triangle FPB$ and, as they are congruent, also of $\triangle FAB$; the latter is but R_A. As the diameter of this circle, $PX = 2R_A$. Similarly, $QY = 2R_C$ and $SZ = 2R_E$. The parallelograms on the diagram show that $PF = a$, $SF = d$, $QB = c$, $PB = f$, $SD = e$, $QD = b$.

Rewriting the claim as

$$2R_A + 2R_C + 2R_E \geq a + b + c + d + e + f$$

and substituting the previous results yields

(2) $$PX + QY + SZ \geq PF + SF + QB + PB + SD + QD;$$

this, as it stands, is a generalization of the celebrated Erdős–Mordell inequality. The actual theorem itself goes like this:

If F, B, D are interior points of the sides ZX, XY, YZ of $\triangle XYZ$, respectively, and the perpendiculars to the respective sides at these points meet at P, Q and S (*Figure 1996/5.2*) then (2) holds.

If the hexagon happens to be centrally symmetric and thus the opposite sides, apart from being parallel, are also equal then $P \equiv Q \equiv S$ and (2) becomes the original Erdős–Mordell inequality.

The solution of the problem hence also proves this generalization; (2), nevertheless, can be demonstrated directly thus yielding an alternative solution ([42]).

1996/6. *Let p, q, n be positive integers with $p + q < n$. Let x_0, x_1, ..., x_n be integers such that $x_0 = x_n = 0$, and for each $1 \leq i \leq n$, $x_i - x_{i-1} = p$ or $-q$. Show that there exist indices $i < j$ with $(i, j) \neq (0, n)$ such that $x_i = x_j$.*

Solution. The *g.c.d.* of p and q divides x_i for every i. This is immediate for x_1 by the initial conditions $x_1 = x_1 - x_0 = p$ or $-q$ and then follows by induction. Dividing through by (p, q) the conditions of the problem still hold and thus $(p, q) = 1$ may be assumed.

Consider the 'telescopic' sum
$$(x_n - x_{n-1}) + (x_{n-1} - x_{n-2}) + \ldots + (x_2 - x_1) + (x_1 - x_0) = 0.$$
If there are k ones among these differences equal to p then the remaining $(n - k)$ ones are $-q$ and thus $pk - q(n - k) = 0$. Since $(p, q) = 1$, q divides k which thus can be written as aq (a is a positive integer). Hence $apq - q(n - aq) = 0$,

(1)
$$n = a(p + q);$$
observe that the condition $n > p + q$ implies $a > 1$.

Consider now the differences
$$y_i = x_{i+p+q} - x_i \qquad (i = 0, 1, \ldots, n - (p+q));$$
it is enough to show that some of them are equal to zero; indeed, if $y_i = 0$ then $x_i = x_{i+p+q}$.

Write down the $(p + q)$-term telescopic sum for y_i:
$$y_i = (x_{i+p+q} - x_{i+p+q-1}) + (x_{i+p+q-1} - x_{i+p+q-2}) + \ldots$$
$$\ldots + (x_{i+2} - x_{i+1}) + (x_{i+1} - x_i).$$
If there are r ones among these differences whose value is p then the remaining $p + q - r$ ones are all equal to $-q$, therefore

(2)
$$y_i = rp - (p + q - r)q = (p + q)(r - q).$$
Hence $(p + q)$ divides every y_i. Consider now the difference $y_{i+1} - y_i$:
$$y_{i+1} - y_i = (x_{i+p+q+1} - x_{i+1}) - (x_{i+p+q} - x_i) = (x_{i+p+q+1} - x_{i+p+q}) - (x_{i+1} - x_i).$$
In the last form each difference is equal to p or $-q$ and thus

$$y_{i+1} - y_i = \begin{cases} 0, \\ \text{or } p + q, \\ \text{or } -(p + q), \end{cases}$$

that is

(3)
$$y_{i+1} - y_i = c(p + q), \qquad \text{where } c = 0, 1 \text{ or } -1.$$

Now the sum
$$(x_{p+q} - x_0) + (x_{2(p+q)} - x_{p+q}) + (x_{3(p+q)} - x_{2(p+q)}) + \ldots + (x_{a(p+q)} - x_{(a-1)(p+q)}) =$$
$$= 0,$$
remembering also that, by (1), $x_n - x_{n-(p+q)} = x_{a(p+q)} - x_{(a-1)(p+q)}$. In terms of the y-variables this means that
$$y_0 + y_{p+q} + y_{2(p+q)} + y_{3(p+q)} + \ldots + y_{n-(p+q)} = 0.$$
Hence it is not possible that the y-s are all of the same sign, in the sequence

(4)
$$y_0, y_1, y_2, \ldots, y_{n-(p+q+1)}, y_{n-(p+q)}$$
either there is a positive term followed by a non positive one or vice versa. Assume, for example, that $y_k > 0$ and $y_{k+1} \leq 0$. Now, by (3), $y_k = c_1(p + q)$ and

$y_{k+1} = c_2(p+q)$, $c_1 > 0$, $c_2 \le 0$, therefore $c_2 - c_1 < 0$. On the other hand, by (2)

$$y_{k+1} - y_k = (c_2 - c_1)(p+q) = -(p+q),$$

and thus $c_2 - c_1 = -1$ that is $c_1 = c_2 + 1$. Then

$$0 < c_2 + 1 \le 1,$$

and, finally, we are at $c_2 = 0$ that is $y_{k+1} = 0$, indeed.

The same argument works if there is a negative term in (4) preceding a non negative one.

Remarks. 1. The following sequence, in fact, has the given property if $n = 15$, $p = 3$, $q = 2$ and $a = 3$:

$$0, \ 3, \ 6, \ 9, \ 7, \quad 5, \ 8, \ 11, \ 9, \ 7, \quad 10, \ 8, \ 6, \ 4, \ 2, \quad 0.$$

2. Having checked the solution one can — and should — ask that where did the idea of the sequence y_i come from. This seems to be quite a rabbit out of the hat, however, if one attempts to prepare an actual sequence of the problem and remembers also the hard to miss property (1) then the $p + q$ neighbouring terms might come to sight as potential candidates for being equal.

1997.

1997/1. *In the plane the points with integer coordinates are the vertices of unit squares. The squares are coloured alternately black and white as on a chessboard. For any pair of positive integers m and n, consider a right-angled triangle whose vertices have integer coordinates and whose legs, of lengths m and n, lie along the edges of the squares. Let S_1 be the total area of the black part of the triangle, and S_2 be the total area of the white part. Let*

$$f(m, n) = |S_1 - S_2|.$$

(a) *Calculate $f(m, n)$ for all positive integers which are either both even or both odd.*

(b) *Prove that $f(m, n) \le \dfrac{1}{2} \max(m, n)$ for all m, n.*

(c) *Show that there is no constant C such that $f(m, n) < C$ for all (m, n).*

Solution. (a) Denote, for brevity, the black and white areas covered by a polygon $XYZU\ldots$ by $S_1(XYZU\ldots)$ and $S_2(XYZU\ldots)$, respectively. Let $AB = m$, $AC = n$ and, by reflecting it to the midpoint F of the hypotenuse, complete the right triangle ABC to the rectangle $ABDC$. If both legs are even then F itself is a lattice point and if the legs are both odd then F is the centre of a lattice square. Therefore, the reflection through F preserves the colouration, any square and its image have the same colour (*Figure 1997/1.1*).

$$S_1(ABC) = S_1(DCB) \qquad \text{and} \qquad S_2(\dot{A}BC) = S_2(DCB).$$

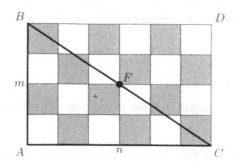 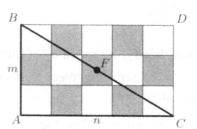

Figure 97/1.1

Hence

$$S_1(ABC) = \frac{1}{2}S_1(ABDC) \quad \text{and} \quad S_2(ABC) = \frac{1}{2}S_1(ABDC).$$

Thus

$$f(m, n) = |S_1(ABC) - S_2(ABC)| = \frac{1}{2}|S_1(ABDC) - S_2(ABDC)|.$$

If both m and n are even then there are the same number of black and white squares contained in the rectangle; if both of them are odd then the **area of the** rectangle is also odd, therefore one colour is ahead of the other one by a single unit. Thus

$$f(m, n) = \begin{cases} 0, & \text{if } m \text{ and } n \text{ are both even and} \\ \dfrac{1}{2}, & \text{if } m \text{ and } n \text{ are both odd.} \end{cases}$$

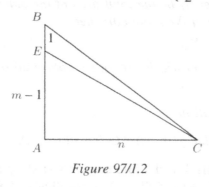

Figure 97/1.2

(b) If $n \equiv m \pmod 2$ then **the** claim follows from (a) because both m **and** n **are** at least 1.

Consider now, for example, when n is even and the other leg, m **is odd.** Choose point E on AB such **that** $BE = 1$ (*Figure 1997/1.2*). Now clearly $EA = m - 1$ and, since both $m - 1$ and n are even, $f(m - 1, n) = 0$. Besides

$$S_1(ABC) = S_1(AEC) + S_1(EBC),$$
$$S_2(ABC) = S_2(AEC) + S_2(EBC).$$

Hence

$$f(m, n) = |S_1(ABC) - S_2(ABC)| = |S_1(EBC) - S_2(EBC)| \leq$$

$$\leq [EBC] = \frac{n}{2} \leq \frac{1}{2}\max(m, n),$$

the desired result.

(c) What we are to prove, actually, is that $f(m, n)$ admits arbitrarily large values, it is unbounded. We show that, in fact, that is the case if the legs are consecutive integers, $m = 2k + 1$ and $n = 2k$.

Consider, as in part (b), the point E on the leg AB for which $BE = 1$; then $AC = AE = 2k$ and

$$f(2k + 1, 2k) = |S_1(EBC) - S_2(EBC)|.$$

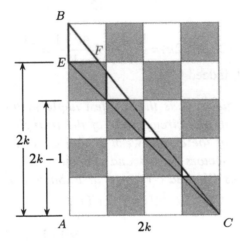

Figure 97/1.3

Observe now the diagram; we clearly may assume that CE, the hypotenuse of the isosceles right triangle ACE is passing through black fields. Calculate now the total white area inside the scalene triangle EBC. The white $\triangle BEF$ as well as all the smaller and smaller white triangles inside $\triangle EBC$ are similar to $\triangle ABC$. $[BEF] = \dfrac{k}{2k + 1}$ because $BE = 1$ and $EF = \dfrac{2k}{2k + 1}$. The subsequent white triangles can be obtained from $\triangle BEF$ by reduction by

$$\frac{2k - 1}{2k}, \ \frac{2k - 2}{2k}, \ \frac{2k - 3}{2k}, \ \dots, \ \frac{1}{2k}$$

respectively and hence, for their areas, $[BEF]$ should be scaled up by the squares of these numbers:

$$\left(\frac{2k - 1}{2k}\right)^2, \ \left(\frac{2k - 2}{2k}\right)^2, \ \left(\frac{2k - 3}{2k}\right)^2, \ \dots, \ \left(\frac{1}{2k}\right)^2.$$

Therefore, the total white area is

$$S_2(EBC) = \frac{k}{2k+1}\left(1 + \left(\frac{2k-1}{2k}\right)^2 + \left(\frac{2k-2}{2k}\right)^2 + \ldots + \left(\frac{1}{2k}\right)^2\right) =$$

$$= \frac{k}{(2k+1)4k^2}\left((2k)^2 + (2k-1)^2 + \ldots + 1^2\right) =$$

$$= \frac{1}{4k(2k+1)}\,\frac{2k(2k+1)(4k+1)}{6} = \frac{4k+1}{12}.$$

Since $[EBC] = k$, its black part is

$$S_1(EBC) = k - \frac{4k+1}{12} = \frac{8k-1}{12}.$$

Hence

$$f(2k+1,\,2k) = \frac{8k-1}{12} - \frac{4k+1}{12} = \frac{2k-1}{6},$$

which is unbounded, indeed.

1997/2. *The angle at A is the smallest angle in the triangle ABC. The points B and C divide the circumcircle of the triangle into two arcs. Let U be an interior point of the arc between B and C which does not contain A. The perpendicular bisectors of AB and AC meet the line AU at V and W, respectively. The lines BV and CW meet at T. Show that*

$$AU = TB + TC.$$

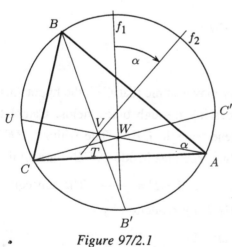

Figure 97/2.1

Solution. Let the lines BV and CW meet the circumcircle at B' and C' (*Figure 1997/2.1*) and denote the perpendicular bisectors of AC and AB by f_1 and f_2, respectively.

The mirror image of CC' through f_1 is AU is mapped to BB' when reflected on through f_2. Hence $CC' = AU = BB'$. Two equal chords in a circle are spanning a symmetrical trapezium and thus $TC = TB'$ yielding

$$AU = BB' = TB + TB' = TB + TC,$$

the claim.

Remark. Since f_1 is taken to f_2 by a rotation of $\angle A$, the composition of the two reflections in the solution is a single rotation by $\angle 2A$. If $\angle A$ is right angle then the resulting rotation is by $180°$ and thus CC' and BB' are parallel or identical; in the first case point T does not exist. The restriction about the magnitude of $\angle A$ is hence to guarantee the existence of T.

1997/3. *Let* x_1, x_2, ..., x_n *be real numbers satisfying*

$$|x_1 + x_2 + \ldots + x_n| = 1$$

and

$$|x_i| \leq \frac{n+1}{2} \qquad (i = 1, 2, \ldots, n).$$

Show that there exists a permutation y_i *of* x_i *such that*

$$|y_1 + 2y_2 + 3y_3 + \ldots + ny_n| \leq \frac{n+1}{2}.$$

Solution. We shall use the well known fact that every permutation can be obtained from any other one by a sequence of *transpositions* that swaps two consecutive elements. Denote the chain of permutations of $x_1, x_2, \ldots, x_n$ leading from $P_0 = (x_1, x_2, \ldots, x_n)$ to the reversed $P_\omega = (x_n, x_{n-1}, \ldots, x_1)$ by $P_0, P_1, \ldots$ $\ldots, P_\omega$; each permutation in the chain is obtained from the previous one by a transposition. For brevity, denote $\frac{n+1}{2}$ by r and for $P_i = (y_1, y_2, \ldots, y_n)$ let

$$S(P_i) = y_1 + 2y_2 + \ldots + ny_n.$$

Let's check now the change of $S(P_i)$. Note that

$$S(P_0) + S(P_\omega) = (x_1 + 2x_2 + \ldots + nx_n) + (x_n + 2x_{n-1} + \ldots + nx_1) =$$
$$= (n+1) \cdot (x_1 + x_2 + \ldots + x_n) = 2r(x_1 + x_2 + \ldots + x_n),$$

and thus

(1) $$|S(P_0) + S(P_\omega)| = 2r.$$

If $|S(P_0)| \leq r$ or $|S(P_\omega)| \leq r$ then we are done; assume, hence, that both inequalities hold the other way round:

$$|S(P_0)| > r \qquad \text{and} \qquad S(P_\omega) > r.$$

This, by (1), means that $S(P_0)$ and $S(P_\omega)$ are of opposite sign:

(2) $$S(P_0) < -r \qquad \text{and} \qquad S(P_\omega) > r,$$

for example. Consider now P_i and P_{i+1}, two consecutive permutations in the chain. Let

$$P_i = (y_1, y_2, \ldots, y_{k-1}, y_k, y_{k+1}, y_{k+2}, \ldots, y_n) \qquad \text{and}$$
$$P_{i+1} = (y_1, y_2, \ldots, y_{k-1}, y_{k+1}, y_k, y_{k+2} \ldots, y_n),$$

with y_k and y_{k+1} being swapped. Here clearly

$$|S(P_{i+1}) - S(P_i)| = |k(y_{k+1} - y_k) + (k+1)(y_k - y_{k+1})| =$$
$$= |y_k - y_{k+1}| \leq |y_k| + |y_{k+1}| \leq 2r.$$

Thus in the sequence $S(P_0), S(P_1), \ldots, S(P_\omega)$ the difference of consecutive terms does not exceed $2r$; $S(P_0)$ and $S(P_\omega)$, on the other hand, are separated by the interval $[-r, r]$ by (2).

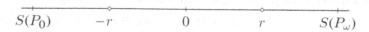

$$S(P_0) \qquad -r \qquad 0 \qquad r \qquad S(P_\omega)$$

Starting from below $-r$ at $S(P_0)$ we hence arrive above r to $S(P_\omega)$ by steps not exceeding $2r$; therefore, the sequence of permutations contains a term P_j such that $S(P_j)$ is in the interval $[-r, r]$ that is

$$|S(P_j)| \le r$$

which completes the proof.

1997/4. *An $n \times n$ matrix whose entries come from the set $S = \{1, 2, \ldots$ $\ldots, 2n-1\}$ is called silver matrix if, for each $i = 1, 2, \ldots, n$, the **ith row and** the ith column together contain all elements of S. Show that:*

(a) *there is no silver matrix for $n = 1997$;*

(b) *silver matrices exist for infinitely many values of n.*

Solution. (a) We show, in general, that there is no silver matrix for n odd, $n > 1$.

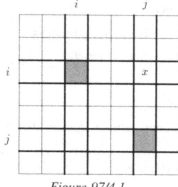

Figure 97/4.1

The union of the kth row and the kth column is going to be called the kth 'cross' ($k = 1, 2, \ldots, n$); each number from S is contained in every cross, by condition. We shall need now an element x of S which does not occur in the main diagonal at all; such a number certainly does exist since $|S| = 2n-1$ and there are only n entries in the main diagonal ($n > 1$).

Consider now such an entry and the two crosses, say the ith one and the jth one passing through this occurrence of x (*Figure 1997/4.1*). The size of a cross is $2n-1$ so this one is the only copy of x in these two crosses, otherwise, some element of S would be missing from one of these two crosses. Consider now any other cross: this, by condition, also contains an x-entry. By the choice of x the other cross through this entry is different from the first one. Coming to the point, there is a matching among the crosses of the matrix, via the occurrences of x and the members of each pair are different. This is the end because, by this matching, the number of crosses, n is even, indeed.

(b) We construct silver matrices of size $n = 2^k$ for every positive integer k. In fact, we show how to prepare a $2n \times 2n$ silver matrix from an $n \times n$ one.

Consider A, a silver matrix of order n. Prepare matrix B by adding $2n$ to each entry and matrix C by replacing the entries in the main diagonal in B by $2n$.

Matrix D is now assembled from these components according to the pattern

$$D = \begin{bmatrix} A & B \\ C & A \end{bmatrix}.$$

We still need a 2×2 silver matrix; A below will do. Then

$$A = \begin{bmatrix} 1 & 2 \\ 3 & 1 \end{bmatrix}, \qquad B = \begin{bmatrix} 5 & 6 \\ 7 & 5 \end{bmatrix}, \qquad C = \begin{bmatrix} 4 & 6 \\ 7 & 4 \end{bmatrix}, \qquad D = \begin{bmatrix} 1 & 2 & 5 & 6 \\ 3 & 1 & 7 & 5 \\ 4 & 6 & 1 & 2 \\ 7 & 4 & 3 & 1 \end{bmatrix}.$$

Any cross in D is contains the numbers $1, 2, \ldots, 2n-1$ in its A-minor and the remaining $2n, 2n+1, \ldots, 4n-1$ values are in B and C; D itself is hence indeed silver.

Remark. The attribute 'silver' is definitely unusual in terms of matrices; the name follows the IMO tradition: whenever there is a chance to do so, the formulation of problems indicate some characteristics of the host country. In Finland or Poland, for example, some points in a problem have been coloured white and blue or white and red, the colours of the national flags, respectively, of these countries. *Argentina*, in latin languages, means silver or rich in silver (the Latin itself for silver is *argentum*).

1997/5. *Find all pairs (a, b) of positive integers that satisfy*

(1) $$a^{b^2} = b^a.$$

Solution. Let a and b be solutions of (1) and denote (a, b) by d. Then $a = du$, $b = dv$, with positive integers u, v such that $(u, v) = 1$. Rewriting (1) as

$$(du)^{d^2 v^2} = (dv)^{du} \qquad \text{yields}$$

(2) $$(du)^{dv^2} = (dv)^u.$$

We shall proceed by distinguishing three cases, namely

(a) $dv^2 = u$, \qquad\qquad (b) $dv^2 > u$, \qquad\qquad (c) $dv^2 < u$.

(a) (2) now becomes

$$(du)^u = (dv)^u, \qquad \text{and thus} \qquad u = v.$$

Since u and v are coprime, $u = v = 1$ and the arising $dv^2 = u$ implies $d = 1$. Therefore $a = b = 1$ and this is clearly a solution.

(b) If $dv^2 > u$ then $dv^2 - u \geq 1$ and now (2) yields

(3) $$d^{dv^2 - u} \cdot u^{dv^2} = v^u.$$

Thus u divides v; using $(u, v) = 1$ again $u = 1$ and (3) hence becomes

(4) $$d^{dv^2 - 1} = v.$$

If $d = 1$ then we get $v = 1$ contradicting $dv^2 - u \geq 1$; if $d \geq 2$ then

$$d^{dv^2 - 1} \geq 2^{2v^2 - 1},$$

which is again impossible; indeed, $2^{2v^2-1} > v$ contradicts (4). Hence we get no solution at all, this time.

(c) The tough part is case (c). Now $dv^2 < u$ and thus $u - dv^2 \geq 1$. This time (2) becomes

$$u^{dv^2} = d^{u-dv^2} \cdot v^u.$$

Now it is v dividing u so, by $(u, v) = 1$ again, $v = 1$ and our equation is

(5)
$$u^d = d^{u-d}.$$

Here, of course, $u > d$ but then $d < u - d$, therefore $u > 2d$.

If p is an arbitrary prime factor of u then, by (5), p also divides d. Let α and δ be the indices of p in u and d, respectively, that is the highest integers for which $p^\alpha \mid u$ and $p^\delta \mid d$. Thus the indices of p on the *l.h.s* and the *r.h.s.* of (5) are αd and $\delta(u - d)$, respectively. Remembering that $u > 2d$, we obtain

$$\alpha d = \delta(u - d) > \delta d,$$

that is $\alpha > \delta$. Hence any prime power divisor of d divides u, as well, the former is a proper divisor of the latter:

(6)
$$u = kd,$$

furthermore, by $u > 2d$, we have $k \geq 3$.

Plug this into (5):

$$(kd)^d = d^{kd-d},$$

(7)
$$k = d^{k-2}.$$

$d = 1$ is not possible since $k \geq 3$. If $k = 3$ then, by (7), we get $d = 3$; $u = 9$ by (6) and hence $a = 27$, finally, $v = 1$ yields $b = 3$. Here there is another solution, the pair (27, 3).

If $k = 4$ then proceeding similarly: (7) yields $d = 2$, hence $u = 8$, by (6), then $a = 16$ and $b = 2$, one more solution.

If $k \geq 5$ then $d^{k-2} \geq 2^{k-2}$, by $d \geq 2$. It is easy to check, by induction, for example that if $k \geq 5$ then $2^{k-2} > k$ and thus (7) cannot hold any more.

There are three solutions of (1), altogether. They are

$$(1, 1), \qquad (27, 3), \qquad (16, 2).$$

1997/6. *For each positive integer n, let $f(n)$ denote the number of ways of representing n as a sum of powers of 2 with non-negative integer exponents. Representations which differ only in the ordering of their summands are considered to be the same. For example, $f(4) = 4$, because 4 can be represented as*

$$4; \qquad 2+2; \qquad 2+1+1; \qquad 1+1+1+1.$$

Prove that for any integer $n \geq 3$

$$2^{\frac{n^2}{4}} < f(2^n) < 2^{\frac{n^2}{2}}.$$

First solution. Getting started we verify two simple recurrences, both satisfied by the function f.

If n is odd then 1 has to be present in its representations; removing it we get a representation of $n - 1$. Conversely, adding a 1 to any representation of the even $n - 1$ yields a representation of n. This one to one mapping shows that

(1) $$f(2k + 1) = f(2k) \qquad (k = 1, 2, \ldots)$$

If $n = 2k$ is even then, depending on the presence of 1, its representations can be divided into two disjoint groups. Removing 1 from the elements of the first group yields representations of $2k - 1$ and, clearly, all of them can be produced this way. Hence there are $f(2k - 1)$ representations of $n = 2k$ that contain 1. The terms in any representation of the second kind, when halved, yield a unique representation of $n/2 = k$ so there are $f(k)$ of them. Therefore

(2) $$f(2k) = f(2k - 1) + f(k).$$

Comparing (1) and (2) yields

$$f(2k) - f(2k - 2) = f(k),$$

$$f(2k - 2) = f(2k - 1) = f(2k) - f(k) < f(2k) = f(2k + 1).$$

Function f is hence increasing.

Clearly $f(1) = 1$. To make life simpler set also the usual $f(0) = 1$. Rearrange (2) as $f(2k) - f(2k - 2) = f(k)$ and substitute here $k = 1, 2, \ldots, n$, respectively:

$$f(2) - f(0) = f(1),$$

$$f(4) - f(2) = f(2),$$

$$\vdots$$

$$f(2n) - f(2n - 2) = f(n).$$

The internal terms do cancel as these differences are added and we obtain

(3) $f(2n) = f(0) + f(1) + \ldots + f(n)$ (n is an arbitrary natural number).

Remember now that f is increasing:

$$f(2n) = 2 + (f(2) + f(3) + \ldots + f(n)) \leq 2 + (n - 1)f(n),$$

(4) $f(2n) \leq nf(n).$

Starting with $n = 2^{n-1}$ apply the above estimation to the decreasing powers: $2^{n-1}, 2^{n-2}, \ldots, 2^1$:

$$f(2^n) \leq 2^{n-1} f\left(2^{n-1}\right) \leq 2^{n-1} \cdot 2^{n-2} f(2^{n-2}) \leq \ldots \leq$$

$$\leq 2^{(n-1)+(n-2)+\ldots+1} f(2) = 2^{\frac{n(n-1)}{2}} \cdot 2 = 2^{\frac{n^2-n+2}{2}}.$$

Since $n^2 - n + 2 < n^2$ if $n \geq 3$,

$$f(2^n) < 2^{\frac{n^2}{2}},$$

indeed, the upper estimation of the problem is hence settled.

To prove the other one note first that if $a \equiv b$ (mod 2) and $b \geq a \geq 0$ then

(5) $$f(b+1) - f(b) \geq f(a+1) - f(a).$$

This is obvious if both a and b are even because then, by (1), both sides are equal to zero. If a and b are both odd then rearranging (2) as $f(2k) - f(2k-1) = f(k)$ and plugging here $b = 2k - 1$ and $a = 2k - 1$, respectively, yields

$$f(b+1) - f(b) = f\left(\frac{b+1}{2}\right), \qquad f(a+1) - f(a) = f\left(\frac{a+1}{2}\right).$$

Since $b + 1 \geq a + 1$ the monotonity of f implies (5).

Let r be now even and apply (5) k times, to the pairs

$$(r, r), \ (r-1, r+1), \ \ldots, \ (r-k+1, r+k-1).$$

Thus

$$f(r+1) - f(r) \geq f(r+1) - f(r),$$
$$f(r+2) - f(r+1) \geq f(r) - f(r-1),$$

$$\vdots$$

$$f(r+k) - f(r+k-1) \geq f(r-k+2) - f(r-k+1).$$

When the resulting inequalities are added the internal terms cancel again yielding

$$f(r+k) - f(r) \geq f(r+1) - f(r-k+1).$$

Remembering that r is odd and thus, by (1), $f(r+1) = f(r)$, the above inequality can also be written as

$$f(r+k) + f(r-k+1) \geq 2f(r) \qquad (k = 1, 2, \ldots, r)$$

Writing this down for the possible values of k

$$f(r+1) + f(r) \leq 2f(r),$$
$$f(r+2) + f(r-1) \leq 2f(r),$$

$$\vdots$$

$$f(r+r) + f(1) \leq 2f(r)$$

and summing once more yields

$$f(1) + f(2) + \ldots + f(2r) \geq 2rf(r).$$

The *l.h.s.*, by (3), is equal to $f(4r) - f(0) = f(4r) - 1$, therefore

$$f(4r) - 1 \geq 2rf(r),$$

that is

(6) $$f(4r) > 2rf(r), \qquad \text{if } r \geq 2 \text{ is even.}$$

For the powers of 2 this yields

(7) $$f(2^m) > 2^{m-1}f(2^{m-2}).$$

(We have set $r = 2^{m-2}$ ($m \geq 3$) in (6)) It is obvious that (7) also holds if $m = 2$: then $f(4) = 4 \geq 2f(0) = 2$.

After this long preparation we finally take off to prove the lower estimation. The main weapon is (7) and it is applied to the values $m = n$, $n - 1$, ..., $n - 2s + 2$ ($n \geq 2s$), respectively:

$$f(2^n) \geq 2^{n-1} f\left(2^{n-2}\right) > 2^{n-1} \cdot 2^{n-3} f\left(2^{n-4}\right) > \ldots$$
$$> 2^{(n-1)+(n-2)+\ldots+(n-2s+1)} f\left(2^{n-2s}\right) = 2^{s(n-s)} \cdot f\left(2^{n-2s}\right).$$

If n is even then set $s = \dfrac{n}{2} - t$:

$$f(2^n) > 2^{\frac{n}{2} \cdot \frac{n}{2}} f\left(2^0\right) = 2^{\frac{n^2}{4}},$$

otherwise let $s = \dfrac{n-1}{2}$:

$$f(2^n) > 2^{\frac{n-1}{2} \cdot \frac{n+1}{2}} f\left(2^1\right) = 2^{\frac{n^2-1}{4}} \cdot 2 = 2^{\frac{n^2+3}{4}} > 2^{\frac{n^2}{4}};$$

the lower estimation is hence proved. (It is, by the way, obviously true for $n = 1$.)

Second solution. In what follows you can see a more direct approach to the upper estimation. Consider a representation of 2^n:

(8) $$2^n = a_0 \cdot 2^0 + a_1 \cdot 2^1 + \ldots + a_n \cdot 2^n,$$

where a_i is non negative integer. A particular solution, for $j = 0, 1, 2, \ldots, n$, is $a_j = 2^{n-j}$ and $a_i = 0$ for $i \neq j$. Hence we obtain $n + 1$ trivial solutions of (8). Note that these solutions are all for the very same, single term representation of 2^n. We now write down an upper bound for the number of non trivial solutions of (8) which, of course, provides an upper bound to the actual number of representations. This multiple counting, by the way, yields a rather poor upper bound, luckily enough it still works.

a_1 can admit the values $0, 1, 2, \ldots, 2^{n-1} - 1$; 2^{n-1} possibilities;

a_2 can admit the values $0, 1, 2, \ldots, 2^{n-2} - 1$; 2^{n-2} possibilities;

$\vdots$

a_{n-2} can admit the values $0, 1, 2, 3 = 2^2 - 1$; 2^2 possibilities;

a_{n-1} can admit the values $0, 1 = 2^1 - 1$; 2^1 possibilities.

Observe that once these values have been set, the value of a_0 is already determined and thus a 2^n factor can be saved.

We don't care about overlaps, this brutal estimate for the number of solutions — which does count each actual representation several times — will already do for our purpose. Indeed

$$f(n) < n+1+2 \cdot 2^2 \cdot \ldots \cdot 2^{n-1} = n+1+2^{\frac{n(n-1)}{2}} = n+1+2^{\frac{n^2}{2}-\frac{n}{2}}.$$

If $n \geq 3$ then this can be weakened on:

$$f(n) < n+1+2^{\frac{n^2}{2}-1} = \frac{1}{2}(2n+2)+\frac{1}{2}2^{\frac{n^2}{2}}.$$

For $n=3$ clearly $2n+2 < 2^{\frac{n^2}{2}}$ because 2^3 is already less than $2^{4.5}$ and from here onwards $2n+2$ is growing linearly versus the exponential growth of $2^{\frac{n^2}{2}}$. Being so

$$f(n) < \frac{1}{2} \cdot 2^{\frac{n^2}{2}} + \frac{1}{2} \cdot 2^{\frac{n^2}{2}} = 2^{\frac{n^2}{2}},$$

indeed.

    For the lower estimation we need some bits from the first solution. We start by showing that f is *weakly convex* on the set of even numbers. This means that

(9) $2f(2k) \leq f(2k-2l) + f(2k+2l) \qquad (l \leq k).$

To prove this property it is enough to show, by (3), that

$$2\left(f(0)+f(1)+\ldots+f(k-l)+\ldots+f(k)\right) \leq \left(f(0)+f(1)+\ldots+f(k-l)\right)+$$
$$+\left(f(0)+f(1)+\ldots+f(k-l)+\ldots+f(k)+\ldots+f(k+l)\right),$$

$$f(k-l+1)+f(k-l+2)+\ldots+f(k) \leq f(k+1)+f(k+2)+\ldots+f(k+l),$$

which clearly holds, term by term, since f is increasing (as it has been proved in the first solution).

    The lower estimate now can be proved by induction on n. $f\left(2^1\right) = 2 > 2^{\frac{1}{4}}$, $f\left(2^2\right) = 4 > 2^{\frac{4}{4}} = 2$, the claim hence holds for $n=1,\ 2$. Let $n > 2$ and assume that

(10) $f\left(2^{n-2}\right) > 2^{\frac{(n-2)^2}{4}};$

we show that it follows also for n.

    Consider the recurrence (3) for $f\left(2^n\right)$ and replace the values admitted at odd arguments according to (1):

$$f\left(2^n\right) =$$
$$= f\left(2^{n-1}\right) + f\left(2^{n-1}-1\right) + f\left(2^{n-1}-2\right) + f\left(2^{n-1}-3\right) + \ldots + f(1) + f(0) =$$
$$= f\left(2^{n-1}\right) + 2f\left(2^{n-1}-2\right) + 2f\left(2^{n-1}-4\right) + \ldots$$
$$\ldots + 2f\left(2^{n-1} - \left(2^{n-2}-2\right)\right) + 2f\left(2^{n-1}-2^{n-2}\right) + \ldots$$
$$\ldots + 2f(2) + 2f(0) =$$

$$= \left[f\left(2^{n-1}\right) + f(0) \right] + 2\left[f\left(2^{n-1}-2\right) + f(2) \right] + 2\left[f\left(2^{n-1}-4\right) + f(4) \right] + \ldots$$
$$\ldots + 2\left[f\left(2^{n-1} - \left(2^{n-2}-2\right)\right) + f\left(2^{n-2}-2\right) \right] + 2f\left(2^{n-2}\right) + f(0).$$

Apply now (9) for the 2^{n-3} terms in the brackets:

$$f\left(2^n\right) \geq 2f\left(2^{n-2}\right) + 4f\left(2^{n-2}\right) + \ldots + 4f\left(2^{n-2}\right) + 2f\left(2^{n-2}\right) + 1 >$$
$$> 4 \cdot 2^{n-3} f\left(2^{n-2}\right).$$

The *r.h.s.* can be estimated, by the induction hypothesis, from below:

$$f\left(2^n\right) > 4 \cdot 2^{n-3} \cdot 2^{\frac{(n-2)^2}{4}} = 2^{n-1+\frac{n^2}{4}-n+1} = 2^{\frac{n^2}{4}},$$

the proof is hence finished.

1998.

1998/1. *In the convex quadrilateral $ABCD$, the diagonals AC and BD are perpendicular and the opposite sides AB and CD are not parallel. The point P, where the perpendicular bisectors of AB and CD meet, is inside $ABCD$. Prove that $ABCD$ is cyclic if and only if the triangles ABP and CDP have equal areas.*

Solution. Place the quadrilateral in the Cartesian system in such a way that the diagonals AC and BD are lying along the x and y axes, respectively; they hence meet at the origin O. (*Figure 1998/1.1*). The vertices are $A(0, a)$, $B(b, 0)$, $C(0, c)$ and $D(d, 0)$; the perpendicular bisectors of the opposite sides AB and CD meet at $P(x, y)$, and, accordingly, the feet of the perpendiculars from P to the diagonals are $T(x, 0)$ and $S(0, y)$, respectively. In what follows we shall be working with signed areas.

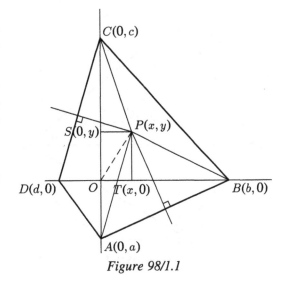

Figure 98/1.1

Note first that

$$2[ABP] = 2[ABO] - 2[APO] - 2[BPO] = ab - ax - by \qquad \text{and}$$
$$2[CDP] = 2[CDO] - 2[CPO] - 2[DPO] = cd - cx - dy.$$

Observe that the two areas $[ABP]$ and $[CDP]$ have been computed with the same sign. For their difference

(1) $$2([ABP]-[CDP])=(a-y)(b-x)-(c-y)(d-x).$$

Turning to the proof, assume first that $ABCD$ is a cyclic quadrilateral; then P is the incentre and thus S and T are bisecting the respective diagonals:

$$x=\frac{b+d}{2},\qquad y=\frac{a+c}{2}.$$

Substituting these results to the *r.h.s.* of (1) yields

$$2([ABP]-[CDP])=\frac{1}{4}[(a-c)(b-d)-(c-a)(d-b)]=0,$$

$[ABP]=[CDP]$, indeed.

Assume now, for the converse, that $[ABP]=[CDP]$. Hence, by (1),

(2) $$|a-y||b-x|=|c-y||d-x|.$$

We show that P is equidistant from the vertices of the quadrilateral. Since $PA==PB$ and $PC=PD$, it is enough to show that $PA=PC$. Assume the contrary, for example $PA>PC$ and, at the same time, $PB>PD$. Checking the (maybe degenerate) right triangles CSP, ASP and BTP, DTP the indirect assumption yields $SA>SC$ and also $TB>TD$ or, in terms of the respective coordinates:

$$|a-y|>|c-y|\quad \text{and}|b-x|>|d-x|,$$

contradicting to (2). A similar contradiction follows from the assumption $PA<<PC$. Therefore, PA and PC are indeed equal, $ABCD$ is cyclic.

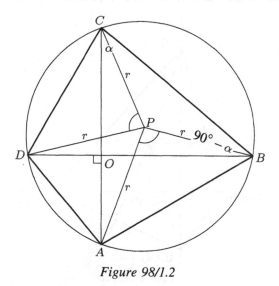

Figure 98/1.2

Remarks. 1. Here there is a simple proof of the first part that in a cyclic quadrilateral $[ABP]==[CDP]$ (*Figure 1998/1.2*). P is the centre of a circle with radius r about O. With the notation $\angle BCO=\alpha$ clearly $\angle CBO==90°-\alpha$, furthermore, as central angles, $\angle APB=2\alpha$ and $\angle CPD==180°-2\alpha$. Thus

$$[ABP]=\frac{r^2}{2}\sin 2\alpha=$$

$$=\frac{r^2}{2}\sin(180°-2\alpha)=[CDP].$$

2. There are various ways to solve the problem; there is, however, a technical difficulty: as P varies in the triangular regions which the quadrilateral is

divided into by the diagonals, the area relations have to be updated and verified over and over again. The tedious case checking can be bypassed via the analytic approach.

In what follows an alternative, still analytic solution will be sketched and thus we don't have to worry about the position of P.

The equations of the perpendicular bisectors of the sides AB and CD are (with the notations of *Figure 1998/1.1*):

$$2bx - 2ay = b^2 - a^2,$$
$$2dx - 2cy = d^2 - c^2.$$

Solving this simultaneous system yields the coordinates of the intersection $P(x_0, y_0)$:

$$x_0 = \frac{c(b^2 - a^2) - a(d^2 - c^2)}{2(bc - ad)} \quad \text{and} \quad y_0 = \frac{d(b^2 - a^2) - b(d^2 - c^2)}{2(bc - ad)}.$$

Since P is interior to the quadrilateral, $\triangle ABP$ and $\triangle CDP$ are oriented similarly. The well known analytic area formulas

$$2[ABP] = ax_0 + by_0 - ab \quad \text{and} \quad 2[CDP] = cx_0 + dy_0 - cd$$

hence compute the respective areas with the same sign. These areas are equal if and only if

$$(a - c)x_0 + (b - d)y_0 + cd - ab = 0.$$

Substituting here x_0 and y_0 as computed above yields

$$(ac - bd)\left[(a - c)^2 + (b - d)^2\right] = 0.$$

The term in the brackets is not zero because a and c, as well as b and d are of opposite sign. Hence $ac - bd = 0$ and thus

$$|a||c| = |b||d|, \quad \text{that is} \quad OA \cdot OC = OB \cdot OD,$$

yielding that A, B, C and D are lying on a circle, indeed.

1998/2. *In a competition there are a contestants and b judges, where $b \geq 3$ is an odd integer. Each judge rates each contestant as either "pass" or "fail". Suppose k is a number such that for any two judges their ratings coincide for at most k contestants. Prove*

$$\frac{k}{a} \geq \frac{b - 1}{2b}.$$

Solution The problem can be represented in a *bipartite graph*. Assign a vertex to every contestant and also to every judge and denote these vertices by $v_1, v_2, \ldots, v_a$ and $w_1, w_2, \ldots, w_b$, respectively. Vertex v is connected to vertex w by a green edge if contestant v has passed in the rating of judge w and in case of fail the connecting edge is red. (The corresponding line in the diagram is continuous for green and dotted for red.) To make it simpler a person and a corresponding vertex will not be distinguished.

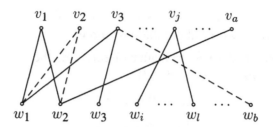

Since every judge has rated all the contestants, the degree of the vertices w_i is a and that of the vertices v_j is b. A pair of similarly coloured edges will be called *hook* if it connects two w vertices via a v vertex. Any hook in the graph means that contestant v has been rated similarly by the corresponding two judges. There are $\binom{b}{2}$ pairs formed from the judges and since, by condition, there are at most k hooks connecting any two judges, the maximal number of hooks is $k\binom{b}{2}$.

Let us count now the hooks from the contestants' side. If there are g_i green edges and r_i red ones incident to the vertex v_i then clearly $g_i + r_i = b$ and there are $\binom{g_i}{2} + \binom{r_i}{2}$ pairs of the same colour through this vertex. Therefore, there are

$$h_i = \binom{g_i}{2} + \binom{r_i}{2}$$

hooks at v_i. On the other hand

$$h_i = \frac{g_i(g_i - 1)}{2} + \frac{r_i(r_i - 1)}{2} = \frac{1}{2}\left(g_i^2 + r_i^2 - (g_i + r_i)\right) =$$

$$= \frac{g_i^2 + r_i^2}{2} - \frac{b}{2} \geq \left(\frac{g_i + r_i}{2}\right)^2 - \frac{b}{2} = \frac{b^2}{4} - \frac{b}{2} = \frac{(b-1)^2}{4} - \frac{1}{4}.$$

Here h_i is integer and since b is odd, $\frac{(b-1)^2}{4}$ is also an integer. Therefore we have the stronger $s_i \geq \frac{(b-1)^2}{4}$. The number of hooks from the contestants side is hence $\sum_{i=1}^{a} h_i \geq \dfrac{a(b-1)^2}{4}$. Comparing this with the previous estimation yields

$$k\binom{b}{2} \geq \frac{a(b-1)^2}{4},$$

$$\frac{k}{a} \geq \frac{b-1}{2b},$$

indeed.

1998/3. *For any positive integer n, let $d(n)$ denote the number of positive divisors of n (including 1 and n). Determine all positive integers k such that*

$$\frac{d(n^2)}{d(n)} = k$$

for some n.

Solution. Note first the well known property of the function $d(n)$: if the prime factorisation of n is $n = p_1^{\alpha_1} p_2^{\alpha_2} \ldots p_r^{\alpha_r}$ then

$$d(n) = (\alpha_1 + 1)(\alpha_2 + a) \ldots (\alpha_r + 1);$$

from here already follows that $d(n)$ is *multiplicative*: if a is prime to b then $d(ab) = d(a) \cdot d(b)$; this, of course, holds for any number of factors (see [44]). Hence

(1) $$\frac{d(n^2)}{d(n)} = \frac{2\alpha_1 + 1}{\alpha_1 + 1} \cdot \frac{2\alpha_2 + 1}{\alpha_2 + 1} \ldots \frac{2\alpha_r + 1}{\alpha_r + 1} = k.$$

Observe that the numbers k we are looking for are the ones that can be written in the form (1) because the corresponding numbers $\alpha_1, \alpha_2, \ldots, \alpha_r$ can be then combined arbitrarily with distinct primes $p_1, p_2, \ldots, p_r$ yielding k as the value of $d(n^2)/d(n)$.

Note, first of all, that, by (1), k has to be odd. Checking a few small odd values with bare hands is quite reassuring: they can indeed be written as required:

$$\frac{d(1^2)}{d(1)} = 1, \qquad \frac{d\left((2^4 \cdot 3^2)^2\right)}{d(2^4 \cdot 3^2)} = \frac{9 \cdot 5}{5 \cdot 3} = 3, \qquad \ldots$$

There seems to be no obvious obstacle and the opposite would be quite awful to prove so let's set out to prove that, indeed, k can admit any odd value. Induction seems to be the right strategy and the initial step has already been completed. Let $k > 3$ and assume that for every odd k_0, less than k, there exists n_0 such that

(2) $$\frac{d(n_0^2)}{d(n_0)} = k_0 \geq 1.$$

We are to prove that k can also be written as desired.

Let's try to set the corresponding indices $\alpha_1, \alpha_2, \ldots, \alpha_r$ in such a way that in the fraction form of k on the *r.h.s.* of (1) we should be able to cancel as much as possible.

As an even number, $k+1$ can be written as $2^r k_0$ where $k_0 \geq 1$ is an odd number and r is a positive integer. Since $k_0 < k$, it can be written in the form (2) by the induction hypothesis. Let

$$\alpha_1 = (2^r - 1)k_0 - 1 \quad \text{and} \quad \alpha_{i+1} = 2^i \alpha_i \qquad (i = 1, 2, \ldots, r)$$

Choose now the primes $p_1, p_2, \ldots, p_r$ all coprime to n_0 and set

$$n = p_1^{\alpha_1} p_2^{\alpha_2} \ldots p_r^{\alpha_r} n_0.$$

With this choice

$$\frac{d(n^2)}{d(n)} = \frac{2\alpha_1 + 1}{\alpha_1 + 1} \cdot \frac{2\alpha_2 + 1}{\alpha_2 + 1} \cdots \frac{2\alpha_r + 1}{\alpha_r + 1} \frac{d(n_0^2)}{d(n_0)} =$$

$$= \frac{\alpha_2 + 1}{\alpha_1 + 1} \cdot \frac{\alpha_3 + 1}{\alpha_2 + 1} \cdots \frac{\alpha_{r+1} + 1}{\alpha_r + 1} k_0 = \frac{\alpha_{r+1} + 1}{\alpha_1 + 1} k_0 =$$

$$= \frac{2^r \alpha_1 + 1}{(2^r - 1)k_0} k_0 = \frac{2^r \alpha_1 + 1}{2^r - 1} = \frac{2^r(\alpha_1 + 1)}{2^r - 1} - 1 =$$

$$= \frac{2^r(2^r - 1)k_0}{2^r - 1} - 1 = 2^r k_0 - 1 = k,$$

indeed, the proof is finished.

1998/4. *Determine all pairs (a, b) of positive integers such that $(ab^2 + b + 7)$ divides $(a^2 b + a + b)$.*

Solution Let $A = a^2 b + a + b$ and $B = ab^2 + b + 7$. Observe that if B divides A then it also divides $bA - aB$; we are hence looking for those pairs (a, b) for which B divides

$$bA - aB = b^2 - 7a.$$

Since $a \geq 1$ we have $b^2 \leq ab^2$, therefore $b^2 - 7a < ab^2 + b + 7 = B$.

Let $b^2 - 7a \geq 0$ first. Then B is dividing a non negative integer smaller than B itself and thus the latter, $b^2 - 7a$ is equal to zero. Then $b^2 = 7a$ and thus b, as a multiple of 7, can be written as $b = 7c$; hence $49c^2 - 7a = 0$ and $a = 7c^2$ (c is a positive integer). These pairs $(a, b) = (7c^2, 7c)$ already satisfy the conditions since

$$A = 49c^4 \cdot 7c + 7c^2 + 7c = 7c(49c^4 + c + 1),$$

$$B = 7c^2 \cdot 49c^2 + 7c + 7 = 7(49c^4 + c + 1)$$

that is $A = cB$.

Assume now that $b^2 - 7a < 0$; its opposite, the positive integer $7a - b^2 < 7a$ is divisible by B. Note that b now cannot exceed 2, otherwise $B = ab^2 + b + 7 \geq 9a + 10$ and thus B cannot divide a positive number less than $7a$. Hence $b = 1$ or $b = 2$. The rest is straightforward:

a) if $b = 1$ then $7a - b^2 = 7a - 1$. If this is a multiple of $B = a + 8$ then writing $7a - 1 = 7(a + 8) - 57$ shows that B has to divide 57. The positive divisors of 57 are: 57, 19, 3, 1, and thus $a + 8$ is 19 or 57, $a = 11$ or $a = 49$. These pairs can be now checked:

$$A = 133 = 7 \cdot 19, \qquad B = 19, \quad \text{resp.} \quad A = 2451 = 43 \cdot 57, \qquad B = 57.$$

b) if $b = 2$ then $7a - b^2 = 7a - 4$ and $B = 4a + 9$. The division $4(7a - 4) = 7(4a + 9) - 79$ yields that B is now dividing 79. Since this prime number has no divisor of the form $4a + 9$ we get no solution if $b = 2$.

Coming to the end: the solutions are the pairs $(7c^2, 7c)$, c is arbitrary positive integer and the two extras: (11, 1) and (49, 1).

1998/5. *Let I be the incentre of the triangle ABC. Let the incircle of ABC touch the sides BC, CA, AB at K, L, M, respectively. The line through B parallel to MK meets the lines LM and LK at R and S, respectively. Prove that the angle RIS is acute.*

Solution In the isosceles triangle LAM clearly $\angle AML = \angle BMR\angle = 90° - \dfrac{A}{2}$; similarly: $\angle CKL = \angle BKS = 90° - \dfrac{C}{2}$ and $\angle BKM = \angle BMK = 90° - \dfrac{B}{2}$. Besides we also have $\angle LMK = \angle MRS = 90° - \dfrac{C}{2}$. Applying the sine rule to $\triangle BRM$ and $\triangle BSK$:

$$(1) \qquad \frac{RB}{BM} = \frac{\sin\left(90° - \frac{A}{2}\right)}{\sin\left(90° - \frac{C}{2}\right)} = \frac{\cos\frac{A}{2}}{\cos\frac{C}{2}}; \quad \text{similarly} \quad \frac{SB}{BK} = \frac{\cos\frac{C}{2}}{\cos\frac{A}{2}}.$$

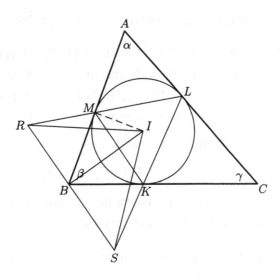

Figure 98/5.1

Hence, by $BK = BM$

(2) $$RB \cdot SB = BM^2.$$

KM and RS are parallel and IB is perpendicular to KM; hence IB is also perpendicular to RS. In the right triangles RBI and SBI

(3) $$IR^2 = RB^2 + IB^2 \qquad \text{and} \qquad IS^2 = SB^2 + IB^2.$$

The cosine rule is ready now in $\triangle RSI$:

$$RS^2 = IR^2 + IS^2 - 2IR \cdot IS \cos \angle RIS.$$

Hence, by (3) and (2) we obtain

$$2IR \cdot IS \cos \angle RIS = IR^2 + IS^2 - RS^2 = RB^2 + SB^2 + 2IB^2 - (RB + SB)^2 =$$
$$= 2IB^2 + RB^2 + SB^2 - RB^2 - SB^2 - 2RB \cdot SB = 2(IB^2 - BM^2).$$

Finally, in the right triangle BIM we have $IB^2 - BM^2 = IM^2$ and hence

$$\cos \angle RIS = \frac{IM^2}{IR \cdot IS},$$

a positive number and thus $\angle RIS$ is acute, indeed.

1998/6. *Consider all functions f from the set of all positive integers into itself satisfying*

(1) $$f\left(t^2 f(s)\right) = s\left(f(t)\right)^2$$

for all s and t. Determine the least possible value of $f(1998)$.

Solution Denote the set of the solutions of (1) by H and let $f \in H$ be an arbitrary solution. Denote $f(1)$ by a. In the first part of the argument we are going to construct an element g of H smaller than f in the usual sense that is its values are smaller (not greater) than the respective values of f.

Setting $t = 1$ in (1) and also $s = 1$ yields

(2) $\quad f(f(s)) = a^2 s;$ (3) $\quad f(at^2) = (f(t))^2 .$

In the computations below the labels above the equality signs are to indicate the actual relation that has been used in that very step.

$$(f(s)f(t))^2 = (f(s))^2 (f(t))^2 \overset{(3)}{=} (f(s))^2 f(at^2) \overset{(1)}{=}$$
$$\overset{(1)}{=} f\left(s^2 f\left(f(at^2)\right)\right) \overset{(2)}{=} f\left(s^2 a^2 \cdot at^2\right) = f\left(a(ast)^2\right) \overset{(3)}{=} (f(ast))^2 .$$

Since the values are positive, here we can write

(4) $$f(s)f(t) = f(ast)$$

and hence, plugging $t = 1$, we obtain $f(as) = af(s)$. Perform here the substitution $s \to st$:

$$f(ast) = af(st),$$

which, by (4), yields

(5) $$af(st) = f(s)f(t).$$

A straightforward induction on k shows that $(f(t))^k = a^{k-1}f(t^k)$. If $k = 1$ then this is the meaningless $f(t) = f(t)$. Let $k > 1$ and assume that $(f(t))^{k-1} = a^{k-2}f(t^{k-1})$. Multiplying by $f(t)$ and applying (5) yields

(6) $$(f(t))^k = a^{k-2}f(t^{k-1})f(t) = a^{k-1}f(t^k),$$

that's it.

Next we prove that a divides $f(t)$ $(t \in \mathbf{N})$. If p is an arbitrary prime and its indices in a and $f(t)$ are α and β, respectively, then $\beta \geq \alpha$ has to be checked.

Clearly $(f(t))^k$ is a multiple of $p^{k\beta}$ and a^{k-1} is that of $p^{(k-1)\alpha}$ for every k. Equality (6) hence yields $k\beta \geq (k-1)\alpha$ and if this holds for every k then $\beta \geq \alpha$, indeed.

Introduce now function g as

$$g(t) = \frac{f(t)}{a}.$$

The values of g, by the previous divisibility relation, are whole numbers, g is mapping $\mathbf{N}$ to itself. Since, by (3), $f(a) = a^2$

(7) $$g(a) = a.$$

Divide now (5) by a^2:

(8) $$g(st) = g(s)g(t);$$

function g is hence *totally multiplicative* and this, of course, can be extended to the product of more arguments. Furthermore (with the previous labelling convention)

$$ag(g(s)) \overset{(7)}{=} g(a)g(g(s)) \overset{(8)}{=} g(ag(s)) = g(f(s)) = \frac{f(f(s))}{a} \overset{(2)}{=} \frac{a^2 s}{a} = as,$$

that is

(9) $$g(g(s)) = s.$$

We are now able to prove that g also belongs to H. The properties (7)–(9) can be used to verify that g indeed satisfies (1):

$$g\left(t^2 g(s)\right) \overset{(8)}{=} g(t^2)g(g(s)) \overset{(8)(9)}{=} s\,(g(t))^2.$$

Thus, for arbitrary $f \in H$ there is a $g \in H$ such that for every $n \in \mathbf{N}$ $g(n) \leq f(n)$. This means that, as far as the problem is concerned, we are free to restrict ourselves to functions satisfying (7)–(9).

Properties (8) and (9) imply that g is one to one and for any prime p the value $g(p)$ is also a prime number. Indeed, by (9)

$$s \overset{(9)}{=} g\,(g(s)) = g\,(g(t)) \overset{(9)}{=} t.$$

For the other property note first that

$$g(1) = \frac{f(1)}{a} = \frac{a}{a} = 1.$$

Let p be now an arbitrary prime and assume that $g(p) = uv$ (u and v are positive integers). By (8) and (9)

$$p = g\,(g(p)) = g(uv) = g(u)g(v),$$

and thus $g(u)$ or $g(v)$ is equal to 1. Since $g(1)$ is also 1 and g is one to one, u or v also has to be equal to 1, that is $g(p)$ is a prime, indeed.

Finally, we can turn to the actual question, the minimal possible value of $f(1998)$. As we have already seen, $f(1998) \geq g(1998)$ and since $1998 = 2 \cdot 3^3 \cdot 37$,

$$g(1998) = g(2) \cdot (g(3))^3 \, g(37).$$

In this product every factor is a prime or the power of a prime. The first primes are 2, 3 and 5, therefore

$$g(1998) \geq 3 \cdot 2^3 \cdot 5 = 120.$$

$f(1998)$ is hence at least 120 for every $f \in H$. We show that there is an element of H — in fact, it will be of the 'g'-kind — such that $g(1998) = 120$. It can/should be defined as follows:

a) $g(1) = 1$;

b) $g(2) = 3$, $g(3) = 2$, $g(5) = 37$, $g(37) = 5$;

c) $g(p) = p$ for every prime p different from 2, 3, 5 and 37;

d) if $n = 2^{\alpha_1} 3^{\alpha_2} 5^{\alpha_3} 37^{\alpha_4} p_5^{\alpha_5} \ldots p_r^{\alpha_r}$, ($\alpha_i = 0$ is possible) then

$$g(n) = (g(2))^{\alpha_1} \, (g(3))^{\alpha_2} \, (g(5))^{\alpha_3} \, (g(37))^{\alpha_4} \, (g(p_5))^{\alpha_5} \ldots (g(p_r))^{\alpha_r}.$$

This function satisfies the conditions (7)–(9) since $g(1) = 1$, (8) holds by the definition, finally

$$g\,(g(n)) = g\left(3^{\alpha_1} \cdot 2^{\alpha_2} \cdot 37^{\alpha_3} \cdot 5^{\alpha_4} \cdot p_5^{\alpha_5} \cdot \ldots \cdot p_r^{\alpha_r}\right) =$$

$$= 2^{\alpha_1} \cdot 3^{\alpha_2} \cdot 5^{\alpha_3} \cdot 37^{\alpha_4} \cdot p_5^{\alpha_5} \cdot \ldots \cdot p_r^{\alpha_r} = n,$$

which is (9). Thus g is in H, indeed, and $g(1998) = 120$, as we have seen, is smaller or equal to $f(1998)$ for every $f \in H$.

1999.

1999/1. *Find all finite sets S of at least three points in the plane such that for all distinct points A, B in S, the perpendicular bisector of AB is an axis of symmetry for S.*

Solution. If $|S| = n$ then the vertices of a regular n-gon clearly satisfy the requirements. We show that there is no other solution, S is the set of vertices of some regular n-gon. If C is the centroid of S then it is also the centroid of any set which is produced by reflecting S in some perpendicular bisector. Therefore C is kept fixed by any symmetry and thus the perpendicular bisectors are all concurrent, they are passing through C. Since reflection preserves distances, this implies that the points of S are all lying on a circle c of centre C.

Consider now A_1, A_2 and A_3, three consecutive points of S, that is A_2 is between A_1 and A_3 and there is no other point of S on the arc $\overline{A_1 A_3}$ *Figure 1999/1.1*. Such a triple certainly exists since the set S is finite.

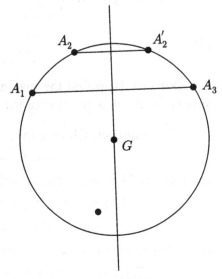

The heart of the issue is that consecutive arcs are equal, $\overline{A_1 A_2} = \overline{A_2 A_3}$. Indeed, if, for example, $\overline{A_1 A_2} \leq \overline{A_2 A_3}$ then A_2', the mirror image of A_2 through the perpendicular bisector of $A_1 A_3$ is interior to the arc $\overline{A_2 A_3}$, a contradiction. The two arcs are equal indeed and thus A_2 is incident to the perpendicular bisector of $A_1 A_3$. We have finished since if any point of S on c is of the same distance from its two neighbours then S forms a regular n-gon inscribed c.

Figure 99/1.1

Remarks. 1. The condition about the finiteness of S was essential; apart from the explicit reference it was also used when the centroid was introduced. If this restriction is released then, besides the points of a circle, there are other solutions, for example, the points of a straight line or the whole plane itself.

2. The original proposal was the more demanding 3D version of the problem: what can be said about the finite set S if it is symmetrical in the perpendicular bisector plane of any two of its points? The answer in this case is that S either forms a regular n-gon (if it is planar) or a platonic solid, a tetrahedron or an octahedron.

1999/2. *Let $n \geq 2$ be a fixed integer. Find the smallest constant C such that for all non-negative reals $x_1, \cdots, x_n$:*

$$(1) \qquad \sum_{1 \leq i < j \leq n} x_i x_j (x_i^2 + x_j^2) \leq C \left(\sum_{1 \leq i \leq n} x_i \right)^4.$$

Determine when equality occurs.

Preliminary remarks. If there is at most one among the given numbers different from zero then the *l.h.s.* of (1) is zero, it holds for any non-negative constant C. In what follows it will be assumed that there are at least two positive ones among the numbers x_i.

In the subsequent solutions the new variables

$$a_i = \frac{x_i}{\displaystyle\sum_{1 \le i < n} x_i}$$

will be used. With these new variables clearly $\Sigma a_i \le 1$ and $0 \le a_i \le 1$ ($i = 1, 2, \ldots$

$\ldots, n$). Dividing (1) by $\left(\displaystyle\sum_{1 \le i \le n} x_i \right)^4$ yields

$$(2) \qquad \sum_{1 \le i < j \le n} a_i a_j (a_i^2 + a_j^2) \le C.$$

Our task is to find the maximum of the *l.h.s.*; its actual value, C, is the answer for the first question of the problem.

First solution. Observe that

$$a_i a_j (a_i^2 + a_j^2) = a_i^3 a_j + a_i a_j^3,$$

and thus the *l.h.s.* of (2) can be written as

$$(a_1 + a_2 + \cdots + a_n)(a_1^3 + a_2^3 + \cdots + a_n^3) - (a_1^4 + a_2^4 + \cdots + a_n^4) =$$
$$= (a_1^3 + a_2^3 + \cdots + a_n^3) - (a_1^4 + a_2^4 + \cdots + a_n^4).$$

Hence the job is to find the maximum of the function

$$F(a_1, a_2, \ldots, a_n) = (a_1^3 + a_2^3 + \cdots + a_n^3) - (a_1^4 + a_2^4 + \cdots + a_n^4)$$

under the condition $\sum a_i = 1$. $1 \ge a_1 \ge a_2 \ge \ldots \ge a_n \ge 0$ can clearly be assumed here. Let a_{k+1} be the last positive one among the numbers a_i in this order; then clearly $k \ge 1$. (The case $k = 1$ will be checked in due course.) Consider now the values of F if the arguments are

$$\mathbf{v}(a_1, a_2, \ldots, a_{k+1}, 0, \ldots, 0) \quad \text{and} \quad \mathbf{v}'(a_1, a_2, \ldots, a_{k-1}, a_k + a_{k+1}, 0, \ldots, 0)$$

respectively. We show that switching from $\mathbf{v}$ to $\mathbf{v}'$ the value of F is increasing; the sum of the coordinates, at the same time, remains the same.

$$F(\mathbf{v}') - F(\mathbf{v}) = \left(a_1^3 + \cdots + a_{k-1}^3 + (a_k + a_{k+1})^3 \right) -$$
$$\left(a_1^4 + \cdots + a_{k-1}^4 + (a_k + a_{k+1})^4 \right) - (a_1^3 + \cdots + a_{k+1}^3) + (a_1^4 + \cdots + a_{k+1}^4) =$$
$$= (a_k + a_{k+1})^3 - a_k^3 - a_{k+1}^3 + a_k^4 + a_{k+1}^4 - (a_k + a_{k+1})^4 =$$
$$= 3a_k a_{k+1}(a_k + a_{k+1}) - 2a_k a_{k+1}(2a_k^2 + 3a_k a_{k+1} + 2a_{k+1}^2) =$$
$$= a_k a_{k+1} \left[(a_k + a_{k+1})(3 - 4(a_k + a_{k+1})) + 2a_k a_{k+1} \right].$$

To show that this difference is indeed positive it is enough to check that

$$a_k + a_{k+1} < 3/4.$$

This is straightforward since $a_1 \geq a_k \geq a_{k+1}$ implies $2a_1 \geq a_k + a_{k+1}$ and thus

$$1 \geq a_1 + a_k + a_{k+1} \geq \frac{a_k + a_{k+1}}{2} + a_k + a_{k+1}.$$

Hence

$$1 \geq \frac{3}{2}(a_k + a_{k+1}) \geq \frac{4}{3}(a_k + a_{k+1})$$

therefore

$$F(\mathbf{v}') \geq F(\mathbf{v}).$$

Starting from an arbitrary argument $\mathbf{v}$ the value of F hence can be increased stepwise until every coordinate, apart from the first two, a_1 and a_2, of the argument is equal to zero. At this point

$$F(a_1, a_2, 0, \ldots, 0) = a_1 a_2 (a_1^2 + a_2^2) = a_1 a_2 (1 - 2a_1 a_2) =$$

$$= -2\left(a_1 a_2 - \frac{1}{4}\right)^4 + \frac{1}{8} \leq \frac{1}{8}.$$

Equality holds if $a_1 a_2 = \frac{1}{4}$ that is $a_1(1 - a_1) = \frac{1}{4}$ and the latter holds if and only if $a_1 = a_2 = \frac{1}{4}$.

The smallest constant C of the problem is hence $\frac{1}{8}$ and equality holds if and only if two of the given numbers x_i are equal and the remaining ones are all zero.

Second solution. As in the previous argument we proceed by analyzing inequality (2). The idea of the proof is now to humbly estimate the terms $a_i^2 + a_j^2$ by the total sum of the squares in the *l.h.s.* of (2). Since

$$a_i^2 + a_j^2 \leq a_1^2 + a_2^2 + \cdots + a_n^2 = Q,$$

(3) $$F = \sum_{1 \leq i < j \leq n} a_i a_j (a_i^2 + a_j^2) \leq Q \sum_{1 \leq i < j \leq n} a_i a_j.$$

Since

$$1 = \left(\sum_{1 \leq i \leq n} a_i\right)^2 = Q + 2 \sum_{1 \leq i < j \leq n} a_i a_j,$$

(3) yields

$$F \leq Q \frac{1 - Q}{2} = \frac{1}{8}[1 - (2Q - 1)^2] \leq \frac{1}{8}.$$

The indicated maximum of F can be attained and this happens only if $Q = \frac{1}{2}$ and for every i, j in (3)

(4) $$a_i a_j (a_i^2 + a_j^2) = a_i a_j Q.$$

Since there are at least two positive ones among the a_i by assumption the ordering of these numbers implies that a_1 and a_2 are certainly positive. Thus, by (4)

$$a_1^2 + a_2^2 = \frac{1}{2}.$$

In case of equality, on the other hand, there can be no more positive ones among the numbers a_i otherwise (4) would imply

$$a_1 a_2 Q = a_1 a_2 (a_1^2 + a_2^2) < a_1 a_2 (a_1^2 + a_2^2 + a_i^2) \leq a_1 a_2 Q,$$

a contradiction. Hence, if there is equality in (2) then $a_i = 0$ if $i > 2$. From here one can conclude as above, in the first solution.

1999/3. *Given an $n \times n$ square board with n even. Two distinct squares of the board are said to be adjacent if they share a common side, **but a square is not adjacent to itself**. Find the minimum number of squares that can be marked so that every square (marked or not) is adjacent to at least one marked square.*

Solution. For simplicity any neighbour of a marked square will be called to be *covered* by the very marked square. The task is then to find the minimum number of marked squares if they cover all the n^2 squares on the board. For this we shall introduce a tricky, labelled colouring of the board.

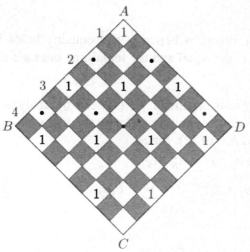

Figure 99/3.1

Colour the fields of the board black and white as if on a chessboard *Figure 1999/3.1*, label the corners and place the board $ABCD$ to have its diagonal BD horizontal. Starting now from the fields adjacent to the side AB enter 1 or · (dot) in the white fields above the horizontal diagonal in the following way: the 'rows' parallel to BD are labelled by the same symbol; every other field of each odd row is labelled by 1 and every other field of each even line is "dotted". Flip now the dotted fields in BD but watch out: the reflected fields below are also labelled by 1.

Let's count now how many squares are labelled by 1, altogether. Their number is clearly the same as the number of fields, 'oned' or dotted, above the horizontal diagonal. Starting from A there are exactly k squares labelled in the k-th white row parallel to BD and thus there are

$$M = 1 + 2 + \ldots + \frac{n}{2} = \frac{n(n+2)}{8}$$

fields oned on the board.

Observe now that each oned square (they are all white) is covered by some black squares but there are no two oned squares such that both of them would be covered by the same black square. Thus we need at least M black squares to cover the white part of the board. The oned squares, on the other hand, do cover the black squares and thus at most M white squares cover the black part.

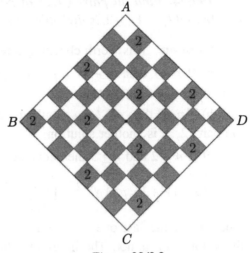

Figure 99/3.2

Now n is even and thus the whole arrangement is symmetric with respect to the two colours. Rotate the board by $90°$ about its centre and label the images of the oned fields by 2 respectively. The 'twoed' fields are all black and they are displayed in *Figure 1999/3.2*.

Swapping the colours temporarily the previous argument now yields an opposite estimate: to cover the black squares we need *at least* M white squares and the white part can be settled by M marked black squares. Coming to the point there are at least

$$N = 2M = \frac{n(n+1)}{4}$$

squares have to be marked and the fields labelled by 1 and 2 when marked show that this many is enough.

Remarks. 1. In a good marking pattern each marked square is adjacent to some other marked square; this can be checked on *Figure 1999/3.3*. There are many other patterns, of course, with the same, minimal number of marked squares.

2. If n is odd then the symmetry of the coloured board is broken; a more tedious argument shows that the required minimum is

$$\left(\frac{n+1}{2}\right)^2 \quad \text{and} \quad \frac{(n+1)^2 - 4}{4}$$

if $n = 4k+1$ and $n = 4k+3$ respectively.

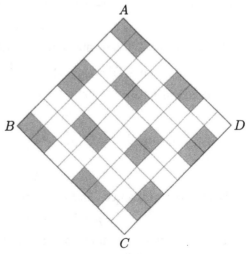

Figure 99/3.3

1999/4. *Find all pairs* (n, p) *of positive integers, such that* p *is a prime,* $n \leq 2p$ *and* $(p-1)^n + 1$ *is divisible by* n^{p-1}.

Solution. It is worth checking a few particular cases first.

a) If $n = 1$ then $(p-1)^1 = p$ is divisible by $1^{p-1} = 1$ and thus the pair $(1, p)$ is a solution for any prime p.

b) If $p = 2$ then $n^{p-1} = n$ should divide $1^n + 1 = 2$ and thus if $n > 1$ then $n = 2$; the pair $(2, 2)$ is another solution.

c) If $n \geq 2$ and $p \geq 3$ then consider again a special case; let $n = p$. Now

$$(p-1)^p + 1 = p^2 \left[p^{p-2} - \binom{p}{1} p^{p-3} + \cdots + \binom{p}{p-3} p - \binom{p}{p-2} + 1 \right].$$

Apart from the last one each term in the brackets is divisible by p and thus the whole sum is not. The index of p hence is at most 2 in the factorisation of $(p-1)^p + 1$ and thus $p - 1 \leq 2$ yielding $p = 3$ as the only possibility. The corresponding pair $(3, 3)$ is another solution.

Note that $n \leq 2p$ holds for every solution so far.

We now show that beyond those already found there is no other solution. Note first that for any further one $n \geq 2$ and $p \geq 3$. Now $(p-1)^n + 1$ is odd and so is n, its divisor by condition. Hence equality cannot hold in the restriction, now $n < 2p$. Consider the smallest prime divisor of n, let it be q. Since q also divides $(p-1)^n + 1$, the two numbers, q and $p - 1$ are coprime. Observe, additionally, that n and $q - 1$ are also coprime since, by q's choice, any proper divisor of n is at least q.

We need a well known piece from the theory of first degree *diophantine equations*: if $(n, q - 1) = 1$ then there exist whole numbers x and y such that

$$nx + (q-1)y = 1.$$

Here $q - 1$ is even and thus x is odd; hence

(1) $$(p-1)^1 = (p-1)^{nx+(q-1)y} = (p-1)^{nx} \cdot (p-1)^{(q-1)y}.$$

The divisibility of $(p-1)^n + 1$ by q can also be written as

$$(p-1)^n \equiv -1 \pmod{q}.$$

On the other hand, since $(q, p-1) = 1$ also holds, *Fermat's theorem* can be invoked as

$$(p-1)^{q-1} \equiv 1 \pmod{q}.$$

These two congruences and (1) imply

$$p - 1 \equiv (-1)^x \cdot 1^y \equiv -1 \pmod{q};$$

thus $p \equiv 0 \pmod{q}$ that is $q > 1$ divides the prime number p. The only possibility is hence $p = q$ so the smallest prime factor of $n < 2p$ is p and thus $n = p$ which was already checked in case c).

The solutions of the problem are hence the pairs $(2,2)$, $(3,3)$ and $(1,p)$ with any prime p.

Remark. Using the assertion of the third problem of the 31st IMO (1990) one can show that there is no other solution even if the restriction $n \leq 2p$ is released.

1999/5. *The circles C_1 and C_2 lie inside the circle C, and are tangent to it at M and N, respectively. C_1 passes through the centre of C_2. The common chord of C_1 and C_2, when extended, meets C at A and B. The lines MA and MB meet C_1 again at E and F. Prove that the line EF is tangent to C_2.*

Solution. We shall use the following

Lemma. If circle c_1 is touching circle c internally at P, a chord AB of c is touching c_1 at Q then F, the point where PQ intersects c bisects the arc $\overgroup{AB}$ (the one not containing P) and, additionally, $FA^2 = FB^2 = FQ \cdot FP$ *(Figure 1999/5.1).*

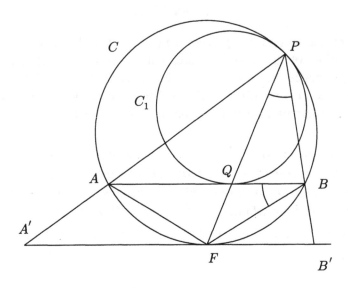

Figure 99/5.1

Putting it differently: given a circle c, a chord AB of it and F bisecting arc AB, c_1 touching c at P and AB at Q, and AB separating c_1 and F, the *power* of F ($FQ \cdot FP$) is constant with respect to any such circle c_1. For the proof of the lemma see the *Remark.*

Let the centres of c_1 and c_2 be O_1 and O_2 respectively and let t_1 and t_2 be the common tangents *(Figure 1999/5.2).*

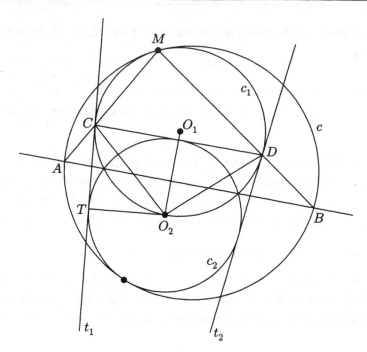

Figure 99/5.2

Now, by the lemma, the powers, with respect to the two circles, c_1 and c_2, of the midpoints of the respective arcs cut by these tangents are equal; these midpoints are hence incident to the radical axis of the two circles. This line is now the common chord of the circles and thus the respective midpoints are but A and B. Our lemma also implies that t_1 and t_2 are touching c_1 at C and D respectively.

Consider now the similarity of centre M which maps c_1 to c. The image of CD, under this enlargement, is clearly AB and thus they are parallel and also perpendicular to O_1O_2. Consequently, O_2 bisects the arc $\overline{CD}$ of c_1 (the one which does not contain M) that is $\angle CO_1O_2 = \angle DO_1O_2$.

Denote the point of contact of t_1 and c_2 by T. By the theorem of inscribed angles in c_1 we get

$$\angle TCO_1 = \frac{1}{2}\angle CO_1O_2 = \frac{1}{2}\angle DO_1O_2 = \angle DCO_2.$$

Hence CO_2 bisects $\angle TCD$, O_2 is equidistant from t_1 and CD and thus CD is touching circle c_2, indeed.

Remark. We now prove the lemma; for the notations please check *Figure 1999/5.1*. Let's see! The similarity by PF/PQ from P maps c_1 to c and, at the same time, AB into the tangent $A'B'$ of c; the equality of the arcs is now straightforward since the point of contact of a tangent parallel to a given chord bisects the arc cut by the chord. Additionally, subtended by equal arcs, $\angle FPB =$ $= \angle QBF$ and thus $\triangle QFB$ and $\triangle BFP$ are similar; this yields $\dfrac{FB}{FQ} = \dfrac{FP}{FB}$ which, when rearranged, becomes $FB^2 = FQ \cdot FP$, indeed.

1999/6. *Determine all functions $f : \mathbf{R} \to \mathbf{R}$ such that*

(1) $\qquad\qquad f\left(x - f(y)\right) = f\left(f(y)\right) + x f(y) + f(x) - 1$

for all x, y in $\mathbf{R}$. [$\mathbf{R}$ is the set of reals.]

First solution. Let $f(0) = c$. Substituting $x = y = 0$ (1) yields
$$f(-c) = f(c) + c - 1.$$
This shows that $c = 0$ is not possible; the contrary would imply $0 = -1$. Choose now an arbitrary element x from $\mathcal{R}_f$, the range of f; if $x = f(y)$ then (1) becomes
$$c = f(x) + x^2 + f(x) - 1,$$

(2)
$$f(x) = \frac{c + 1 - x^2}{2}.$$

We show that f is surjective, $\mathcal{R}_f$ is the set of reals. Setting $y = 0$ in (1)

(3)
$$f(x - c) = f(c) + cx + f(x) - 1,$$
$$f(x - c) - f(x) = cx + f(c) - 1.$$

Let x_0 be arbitrary. Since $c \neq 0$, x can be isolated from
$$cx + f(c) - 1 = x_0.$$
With the solution (3) becomes
$$x_0 = f(x - c) - f(x)$$
and thus any real number can be written as the difference of two elements in $\mathcal{R}_f$. Consider now two arbitrary elements in the range, y_1 and y_2 and let $x_0 = y_1 - y_2$. Setting $x = y_1$ and $f(y) = y_2$ in (1) we get
$$f(x_0) = f(y_1 - y_2) = f(y_2) + y_1 y_2 + f(y_1) - 1.$$
Comparing this with (2) yields
$$f\left(x_0 = \frac{c + 1 - y_2^2}{2} + y_1 y_2 + \frac{c + 1 - y_1^2}{2} - 1 = \frac{2c - (y_1 - y_2)^2}{2} = \frac{2c - x_0^2}{2}\right).$$
Accordingly
$$f(x) = \frac{2c - x^2}{2}$$
for any real number x. Comparing this with (2) again
$$\frac{c + 1 - x^2}{2} = \frac{2c - x^2}{2},$$
that is $c = 1$. Hence
$$f(x) = \frac{2 - x^2}{2}$$
and f can be checked as a solution.

Second solution. Let $f(0) = c$ again and set $x = f(y)$. Then (1) becomes

$$c = f(f(y)) + (f(y))^2 + f(f(y)) - 1,$$

and hence

(4) $$f(f(y)) = \frac{1}{2}\left(c + 1 - (f(y))^2\right).$$

Substitute this to (1) and introduce function g as

$$g(x) = f(x) + \frac{x^2}{2}.$$

Then

$$f(x - f(y)) = \frac{1}{2}\left(c - 1 - (f(y))^2\right) + xf(y) + f(x),$$

$$g(x - f(y)) = f(x - f(y)) + \frac{(x - f(y))^2}{2} = \frac{c - 1 - (f(y))^2}{2} + xf(y) +$$

(5) $$+ g(x) - \frac{x^2}{2} + \frac{(x - f(y))^2}{2} = g(x) + \frac{c - 1}{2}.$$

The constant zero function is not a solution and thus $f(y) \neq 0$ for some y. Plug now $x = \dfrac{1}{f(y)}$ in (1):

(6) $$f(x - f(y)) = f(f(y)) - f(x).$$

Introducing the notations $u = x - f(y)$, $v = f(y)$ and $w = x$ (6) becomes

(7) $$f(u) = f(v) + f(w).$$

Setting $y = u$ in (5) we obtain

$$\underline{g(x - f(u)) = g(x) + c - 1}$$
$$2.$$

Applying (5) twice and using also (7) yields

$$g(x - f(u)) = g[x - f(v) - f(w)] = g(x - f(v)) + \frac{c - 1}{2} = g(x) + c - 1.$$

Comparing these last two equalities we get $\dfrac{c - 1}{2} = c - 1$ that is $c = 1$. Hence, by (5)

$$g(x - f(y)) = g(x)$$

which means that g is periodic and each value of f is a period. Since $f(0) = c = 1$, (4) implies

$$f(f(0)) = f(1) = \frac{1}{2}(1 + 1 - 1)^2 = \frac{1}{2}.$$

Hence $\frac{1}{2}$ is in the range of f and thus it is a period of g. Set now $y=0$ in (1):

$$f(x-1)=f\left(x-f(0)\right)=f\left(f(0)\right)+xf(0)+f(x)-1=\frac{1}{2}+x+f(x)-1=$$

$$=f(x)+x-\frac{1}{2}.$$

Being in the range of f, $f(x)+x-\frac{1}{2}$ is also a period of g which makes altogether three periods found so far: $f(x), \frac{1}{2}$ and $f(x)+x-\frac{1}{2}$. Hence, as their linear combination,

$$x=f(x)+x-\frac{1}{2}-f(x)+\frac{1}{2}$$

is also a period and thus any real number is a period of g. The better for us, function g is constant then. By its definition

$$g(0)=f(0)+\frac{0^2}{2}=1$$

so $g(x)=1$ for every real number x. This is the end; switching back to f we get

$$f(x)=g(x)-\frac{x^2}{2}=1-\frac{x^2}{2}=\frac{2-x^2}{2}$$

and, as substitution shows, this one is satisfying (1) indeed.

2000.

2000/1. *Two circles Γ_1 and Γ_2 intersect at M and N. Let l be the common tangent to Γ_1 and Γ_2 so that M is closer to l than N is. Let l touch Γ_1 at A and Γ_2 at B. Let the line through M parallel to l meet the circle Γ_1 again at C and the circle Γ_2 again at D. Lines CA and DB meet at E; lines AN and CD meet at P; lines BN and CD meet at Q.*

Show that $EP=EQ$.

Solution. Since the point of contact on a tangent to a circle bisects the arc cut by a chord parallel to the tangent line, A and B are the midpoints of the arcs $\overset{\frown}{CM}$ and $\overset{\frown}{DM}$ respectively. Therefore, the triangles ACM and BDM are both isosceles (*Figure 2000/1.1*). Denote the base angles of the triangle ACM by α and those of BDM by β. Angles $EAB\angle$ and $ACM\angle$ are corresponding angles and $BAM\angle$ and $AMC\angle$ are opposite ones. Hence $EAB\angle=BAM\angle=\alpha$ and thus the diagonal AB bisects $A\angle$ of the quadrilateral $EAMB$. Similarly, AB bisects $B\angle$, the quadrilateral $EAMB$ is a deltoid, its diagonal EM is perpendicular to AB and hence also to PQ.

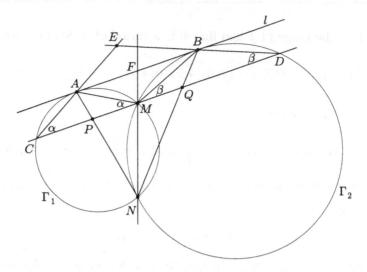

Figure 2000/1.1.

If AB and MN meet at F, then F is incident to the radical axis of the two circles. Therefore, the tangents from F to the two circles are equal, F is the midpoint of segment AB. Since the images of AF and FB under a central similarity from N are PM and MQ respectively, $PM = MQ$.

Therefore, EM is the perpendicular bisector of the segment PQ and thus $EP = EQ$, indeed.

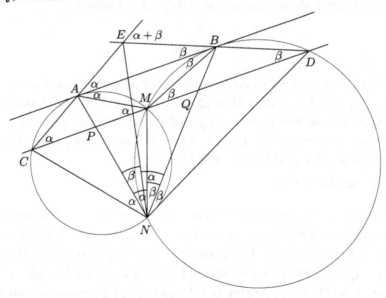

Figure 2000/1.2.

Remark. The proposers suggested an alternative to this problem, too, stating that the very same conditions imply that EN is bisecting $CND\angle$.

The proof now goes as follows: according to what has been already proved (*Figure 2000/1.2*) intercepting the same arcs $CNA\angle = ANM\angle = \alpha$ and also $DNB\angle = BNM\angle = \beta$. Hence

$$CND\angle = 2\alpha + 2\beta \quad \text{and} \quad ANB\angle = \alpha + \beta.$$

Observe that $ANBE$ is cyclic because its external angle at E being also that of the triangle ABE is equal to $\alpha + \beta$, the internal angle of the quadrangle at N. Hence $EAB\angle = ENB\angle = \alpha$ and thus $END\angle = \alpha + \beta$ and this completes the proof.

2000/2. *Let a, b, c be positive real numbers such that $abc = 1$. Prove that*

(1) $$\left(a - 1 + \frac{1}{b}\right)\left(b - 1 + \frac{1}{c}\right)\left(c - 1 + \frac{1}{a}\right) \le 1.$$

First solution. By the condition $abc = 1$ we can apply the following substitutions

$$a = \frac{x}{y}, \quad b = \frac{y}{z}, \quad c = \frac{z}{x}$$

with suitable positive numbers x, y, z (a possible choice is $x = 1$, $y = \dfrac{1}{a}$, $z = \dfrac{1}{ab} = c$). When multiplied by xyz the inequality becomes:

(2) $$(-x + y + z)(x - y + z)(x + y - z) \le xyz.$$

Since the sum of any two factors on the l. h. s. is equal to one of $2x$, $2y$, $2z$, there can be at most one of them less than or equal to zero. If this is the case, however, then the claim is obvious; let's assume now that each factor is positive. There are many ways to verify (2), the argument we chose here is just one posibility. The product of the following obvious inequalities

$$x^2 - (y - z)^2 \le x^2, \quad \text{i. e.} \quad (x + y - z)(x - y + z) \le x^2,$$
$$y^2 - (z - x)^2 \le y^2, \quad \text{i. e.} \quad (y + z - x)(y - z + x) \le y^2,$$
$$z^2 - (x - y)^2 \le z^2, \quad \text{i. e.} \quad (z + x - y)(z - x + y) \le z^2.$$

yields the square of the desired inequality. Since the factors are now positive, (2) follows immediately.

Second solution Let's assume again that none of the three factors is negative. Denote the l. h. s. of (1) by B, and write the terms in each brackets as a single fraction. Their product is a fraction of denominator $abc = 1$ and thus

$$B = (ab - b + 1)(bc - c + 1)(ca - a + 1).$$

Substituting $\dfrac{1}{b} = ac, \dfrac{1}{c} = ab, \dfrac{1}{a} = bc$ in (1) yields

$$B = (a - 1 + ca)(b - 1 + ab)(c - 1 + bc).$$

The product of these two forms of B is

$$B^2 = [(ab - b + 1)(b - 1 + ab)][(bc - c + 1)(c - 1 + bc)][(ca - a + 1)(a - 1 + ca)].$$

Apply now the AM-GM inequality for the expressions inside the square brackets:

$$(ab - b + 1)(b - 1 + ab) \leq \frac{1}{4}(ab - b + 1 + b - 1 + ab)^2 = a^2 b^2,$$

$$(bc - c + 1)(c - 1 + bc) \leq \frac{1}{4}(bc - c + 1 + c - 1 + bc)^2 = b^2 c^2,$$

$$(ca - a + 1)(a - 1 + ca) \leq \frac{1}{4}(ca - a + 1 + a - 1 + ca)^2 = c^2 a^2.$$

Thus $B^2 \leq a^4 b^4 c^4 = 1$, that is $B \leq 1$, and the proof is finished.

2000/3. *Let $n \geq 2$ be a positive integer. Initially, there are n fleas on a horizontal line, not all at the same point.*

For a positive real number λ, define a move *as follows:*

choose any two fleas, at points A and B, with A to the left of B;

let the flea at A jump to the point C on the line to the right of B with $BC/AB = \lambda$.

Determine all the values of λ such that, for any point M on the line and any initial positions of the fleas, there is a finite sequence of moves that will take all the fleas to positions to the right of M.

Solution. First we show that if $\lambda \geq \dfrac{1}{n-1}$ then the fleas can make it to the right of M. Let the pair of fleas selected for each jump be the one to the far left and to the far right, respectively. Consider the pairwise distances after the kth jump and denote the biggest and the smallest one by D_k and d_k, respectively. Clearly $D_k \geq (n-1)d_k$.

After the $(k+1)$th jump there will arise a distance of the size λD_k at the right end. Note that this might be the smallest distance, i. e. $d_{k+1} = \lambda D_k$. In any other case, however, the smallest distance does not decrease, $d_{k+1} \geq d_k$. Accordingly

$$\frac{d_{k+1}}{d_k} \geq \min\left\{1, \frac{\lambda D_k}{d_k}\right\} \geq \min\{1, (n-1)\lambda\} \geq 1.$$

After the first $n-1$ jumps the fleas occupy distinct positions, that is $d_{n-1} > 0$. Since the smallest distance is not decreasing, the flea to the far left is moving steadily to the right from now on, by at least d_{n-1} each time. Therefore, after a while all of them will arrive to the right of M, indeed.

Next we show that if $\lambda < \dfrac{1}{n-1}$ then, regardless of their starting positions, the fleas cannot get arbitrarily far. Consider them now as numbers moving along a number line. Denote, after the kth jump, the sum of their current positions by s_k and the position of the flea to the far right by r_k. Clearly $s_k \leq n \cdot r_k$. We are going to show that no matter how the fleas are jumping around, the sequence r_k remains bounded.

If on the $(k+1)$th round jumping over the flea at B the flea at A is arriving at the point C then denoting the corresponding numbers by a, b and c, respectively, we clearly have $s_{k+1} = s_k + c - a$ and $c - b = \lambda(b-a)$. Writing the latter as $\lambda(c - a) = (1+\lambda)(c-b)$ and combining the two relations yields

$$(1) \qquad s_{k+1} - s_k = c - a = \frac{1+\lambda}{\lambda}(c-b).$$

If $c > r_k$ then flea A has just jumped to the right end, $r_{k+1} = c$. Since $b \leq r_k$, (1) yields

$$(2) \qquad s_{k+1} - s_k = \frac{1+\lambda}{\lambda}(c-b) \geq \frac{1+\lambda}{\lambda}(r_{k+1} - r_k).$$

Observe, that inequality (2) still holds if $c \leq r_k$. Indeed, then $r_{k+1} - r_k = 0$ while $s_{k+1} - s_k = c - a > 0$.

Now $\lambda < \dfrac{1}{n-1}$ and thus $1+\lambda > n\lambda$, that is $\dfrac{1+\lambda}{\lambda} - n > 0$. Introduce $z = \dfrac{1+\lambda}{\lambda} - n$. Rearranging (2):

$$(3) \qquad \frac{1+\lambda}{\lambda} r_k - s_k \geq \frac{1+\lambda}{\lambda} r_{k+1} - s_{k+1}.$$

According to this inequality the sequence $\dfrac{1+\lambda}{\lambda} r_k - s_k$ is not increasing and thus

$$\frac{1+\lambda}{\lambda} r_1 - s_1 \geq \frac{1+\lambda}{\lambda} r_k - s_k.$$

On the other hand $s_k \leq nr_k$ implies

$$\frac{1+\lambda}{\lambda} r_k - s_k = z \cdot r_k + (n \cdot r_k - s_k) \geq z \cdot r_k.$$

Now we have

$$\frac{1+\lambda}{\lambda} r_1 - s_1 \geq z \cdot r_k.$$

Since the sequence $z \cdot r_k$ is bounded, the sequence r_k is also bounded and this was to be proved. The answer to the question of the problem is hence: $\lambda \geq \dfrac{1}{n-1}$.

2000/4. *A magician has one hundred cards numbered 1 to 100. He puts them into three boxes, a red one, a white one and a blue one, so that each box contains at least one card.*

A member of the audience selects two of the three boxes, choses one card from each and announces the sum of the numbers on the chosen cards. Given this sum, the magician identifies the box from which no card has been chosen.

How many ways are there to put all the cards into the boxes so that this trick always works? (Two ways are considered different if at least one card is put into a different box.)

Solution. We are going to prove that there are 12 ways altogether to put the cards in the boxes for this trick to work. For a given arrangement colour each card by the colour of the box containing it, i. e. r, w or b.

Case 1. There are no identically coloured ones in the sequence i, $i+1$, $i+2$ of cards for some i, for example the sequence is rwb. Since $i+(i+3)=(i+1)+(i+2)$, the colour of $i+3$ has to be r, otherwise this very sum would occur in two distinct pairs of boxes. Therefore, three consecutive colours determine the colouring of the subsequent cards, a red number must follow a triple of rwb, then there is a white one and a blue, etc. The argument also works backwards, there is a blue number preceding the rwb sequence and a white one before that, etc.

Therefore, it is sufficient to set the colours of the first three cards 1, 2 and 3; there are 6 different ways to do that. Each periodic arrangement hence obtained is good, since any three sums of the form $r+w$, $w+b$, $b+r$ give different remainders when divided by 3.

Case 2. There are no consecutive triples of distinctly coloured numbers. Assume that number 1 is red. There is a smallest number now that is not red; denote it by i and let it be say white. Consider now the smallest blue number and denote it by t.

Since there is no rwb triple $i+1 < t$. We show that $t = 100$. Assume, to the contrary, that $t < 100$. Since $i+t=(i-1)+(t+1)$ the colour of $(t+1)$ cannot be else but red. Now $i+(t+1)=(i+1)+t$ implies that $(i+1)$ must also be blue, contradicting to the choice of t as the smallest blue number. Accordingly, t must be 100, indeed.

Since $(i-1)+100=i+99$, the colour of 99 must be white. We prove that all the remaining numbers must be white if 1 is red and 100 is the only blue number. Indeed, if some $s > 1$ would be red, then by $s+99=(s-1)+100$ the colour of $(s-1)$ should be also blue which is impossible.

The corresponding colouring is $rww\ldots wwb$ and it yields a good arrangement of the cards. Indeed, if the sum of the selected cards is less than 101 then it is the blue box left out; if the sum is 101 then the white box is not chosen; finally, the red box has been skipped if the sum is exceeding 101. Observe that knowing their sum the magician can even tell the actual cards drawn from respective boxes this time. There are clearly 6 ways again to set the order of the colours.

Remark. This problem was proposed by Hungary, it was set by *Dobos, Sándor*.

2000/5. *Determine whether or not there exists a positive integer n such that n is divisible by exactly 2000 prime numbers, and $2^n + 1$ is divisible by n.*

Solution. Generalizing the question we are going to prove by mathematical induction that there exists a suitable $n(k)$ for arbitrary k, not just for $k = 2000$.

If $k = 1$ then $n(1) = 3$ is clearly a good choice. Let's assume, for $k \geq 1$ that $n(k)$ exists and also that it is of the form $n(k) = 3^l \cdot t$, $l \geq 1$ where $3 \nmid t$, $n(k)$ has exactly k distinct prime divisors and $n(k) \mid 2^{n(k)} + 1$. We are going to show that there is some $n(k+1)$ also satifying the conditions. Since $n = n(k)$ is odd, we have $3 \mid 2^{2n} - 2^n + 1$. Clearly $2^{3n} + 1 = (2^n + 1)(2^{2n} - 2^n + 1)$ and thus $3n \mid 2^{3n} + 1$. According to the *lemma* below there exists some odd prime p satisfying $p \mid 2^{3n} + 1$ and $p \nmid 2^n + 1$. Then $n(k+1) = 3p \cdot n(k)$ is a right choice; the induction is complete.

Lemma: For every integer $a > 2$ there exists a prime p such that $p \mid a^3 + 1$ but $p \nmid a + 1$.

Proof: We argue by reductio ad absurdum. Assume that the claim is false for some integer $a > 2$ and consider an arbitrary prime p divisor of $a^2 - a + 1$. Clearly $p \mid (a^2 - a + 1)(a + 1) = a^3 + 1$ so p divides $(a + 1)$ by assumption. On the other hand

(1) $$a^2 - a + 1 = (a + 1)(a - 2) + 3,$$

and thus p must be equal to 3 forcing $a^2 - a + 1$ to be a power of 3. On the other hand $a - 2 = (a + 1) - 3$ is a multiple of 3. Therefore, in (1), $a^2 - a + 1$ is also divisible by 3. Since it is not a multiple of 9 and it is a power of 3, it has to be equal to 3 itself. However, this is impossible, since $a > 2$ implies $a^2 - a + 1 > 3$.

2000/6. *Let AH_1, BH_2, CH_3 be the altitudes of an acute-angled triangle ABC. The incircle of the triangle ABC touches the sides BC, CA, AB at T_1, T_2, T_3, respectively. Let the lines l_1, l_2, l_3 be the reflections of the lines H_2H_3, H_3H_1, H_1H_2 in the lines T_2T_3, T_3T_1, T_1T_2, respectively.*

Prove that l_1, l_2, l_3 determine a triangle whose vertices lie on the incircle of the triangle ABC.

Solution. The claim essentially states that reflecting the sides of the pedal triangle in the respective sides of the triangle determined by the points of contact of the incircle (the so called contact triangle) yields a triangle (of sides l_1, l_2, l_3) also inscribed the incircle. To get started we prove an auxiliary theorem:

The mirror image of any side of the pedal triangle in the corresponding side of the contact triangle is parallel to a side of the original triangle.

We are going to prove this for the mirror image of the line H_2H_3 in T_2T_3: the line l_1 is parallel to the side BC (*Figure 2000/6.1*). If H_2H_3 and T_2T_3

happen to be parallel or identical then the triangle is isosceles and the claim is obvious. Assume now that, as it is shown on the figure, the lines H_2H_3 and T_2T_3 meet at some point N. The angle of T_2T_3 with both H_2H_3 and l_1 is φ and the intersection of l_1 and AC is Q.

With the compulsory notations $AH_2N\angle = \beta$ because BCH_2H_3 is cyclic. Applying the external angle theorem yields $AT_2N\angle = \varphi + \beta = 90° - \dfrac{\alpha}{2}$, $\varphi = 90° - \dfrac{\alpha}{2} - \beta$ and also

$$AQN\angle = \varphi + AT_2N\angle = 90° - \frac{\alpha}{2} - \beta + 90° - \frac{\alpha}{2} = 180° - (\alpha + \beta) = \gamma.$$

Therefore $ACB\angle = AQN\angle = \gamma$ are corresponding angles and thus l_1 is parallel to BC indeed.

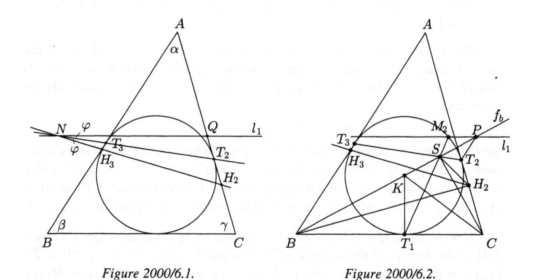

Figure 2000/6.1.   *Figure 2000/6.2.*

Denote the reflections of T_1, T_2, T_3 in the bisectors of $A\angle$, a $B\angle$, a $C\angle$ by M_1, M_2 and M_3, respectively. These points are incident to the incircle since the angle bisectors are its axis of symmetry. We are going to prove that the triangle $M_1M_2M_3$ is identical to the one formed by the lines l_1, l_2, l_3. First we prove that the mirror image of H_2 in the line T_2T_3 is incident to the bisector f_b of the angle $B\angle$. (*Figure 2000/6.2.*). (Similar claim holds for the points H_1, H_3, of course.) Drop, for the proof a perpendicular from H_2 to T_2T_3; let this line meet f_b at P. Denote the intersection of f_b and T_1M_2 by S. It is enough to prove that $PSH_2\angle = 2PST_2\angle$. As opposite angles $PST_2\angle = BST_3\angle$. By the external angle theorem

$$PST_2\angle = BST_3\angle = AT_3S\angle - T_3BS\angle = \left(90° - \frac{\alpha}{2}\right) - \frac{\beta}{2} = 90° - \frac{\alpha}{2} - \frac{\beta}{2} = \frac{\gamma}{2}.$$

On the other hand $BST_1\angle = BST_3\angle$ since they are symmetric with respect to f_b. Denote the incenter by K. Since $KST_1\angle = KCT_1\angle = BST_3\angle = \frac{\gamma}{2}$, the quadrilateral SKT_1C is cyclic because KT_1 subtends the same angle at both S and C. Therefore, the angle $KSC\angle$ opposite to $T_1\angle$ in this quadrilateral is right angle. Hence BCH_2S is also cyclic, since BC subtends a right angle at both H_2 and S. As an external angle of this cyclic quadrilateral $PSH_2\angle = \gamma = 2PST_2\angle$ and this was to be proved.

The fact that BCH_2S is cyclic, also implies that $BPT_2\angle = SH_2T_2\angle = \beta/2$ because they are lying symmetrically with respect to T_2T_3. Since M_2 is the mirror image of T_2 in f_b, $BPM_2\angle = BPT_2\angle = \beta/2$ and thus PM_2 is parallel to BC. As the mirror image of H_2 through T_2T_3 point P is incident to the line l_1. Since l_1 is parallel to BC by our previous result and PM_2 is also parallel to BC this boils down to the conclusion that PM_2 has to be identical to the line l_1, indeed.

If the subscripts 2 and 3 are swapped and vertex B is replaced by C then the previous argument yields that l_1 contains M_3. The very same reasoning for the lines l_2 and l_3 settles the claim.

Remarks. 1. In the preceding considerations we have referred to the positions of certain points, segments and angles as they are given on the *Figure 2000/6.2.* Modifying the shape of the triangle ABC might alter the configuration to some extent; some points may become identical, for example, or lines become parallel, the order of certain points on a line may be different, etc. The proof, however, is essentially valid for all possible configurations. Certain details might need minor modifications of course, instead of a cyclic quadrilateral, for example, the central angle theorem should be applied, etc. Synthetic solutions (i. e. those avoiding the tools of analytic geometry) of such problems necessarily raise this problem. Under the time pressure of the actual contest, however, it is impossible to discuss every possible arrangement and according to general practice these solutions are considered to be complete if the student mentions this issue or states that the solution can be modified without particular difficulties according to the actual configuration.

2. The problem is closely related to Qu.2. of the 1982.IMO. There is the so called Feuerbach theorem in the background in both cases; it states that the incircle and the nine-point circle are touching each other and thus the point of contact is their center of similarity. Enlarging the triangle formed by the midpoints of the sides of triangle ABC by the scale factor $\frac{2r}{R}$ yields the triangle $M_1M_2M_3$ in our problem.

2001.

2001/1. *Let ABC be an acute-angled triangle with circumcentre O. Let P on BC be the foot of the altitude from A. Suppose that $BCA\angle \geq ABC\angle + 30°$. Prove that $CAB\angle + COP\angle < 90°$.*

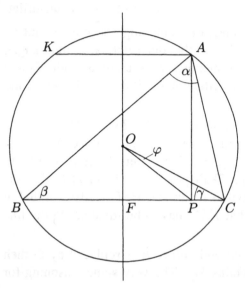

Figure 2001/1.1.

First solution. Denoting the angle $COP\angle$ by φ and using the compulsory notations the condition is $\gamma \geq \beta + 30°$ and we have to prove that $\alpha + \varphi < 90°$ or $\varphi < 90° - \alpha$. If the midpoint of BC is F then clearly $COF\angle = = \alpha$ by the central angle theorem. Since $OCF\angle = OCP\angle = 90° - \alpha$ in the right triangle COF, the claim becomes an inequality for the sides of the triangle $OCP : CP < OP$. (*Figure 2001/1.1.*).

Since $\gamma > \beta$, the point P is lying on the segment FC. The reflection K of A in the perpendicular bisector of BC is also incident to the circumcircle and $KCB\angle = ABC\angle = \beta$. Since $KCA\angle = \gamma - \beta \geq 30°$ the length of the chord intercepted by the angle KCA is exceeding the circumradius: $2R \sin KCA\angle \geq R$. Therefore, $AK \geq R$ and thus

$$FP = \frac{AK}{2} \geq \frac{R}{2}.$$

Since ABC is acute, BC is strictly smaller than the diameter of the circumcircle: $BC < 2R$, and thus $CF < R$. Hence

$$CP = CF - FP < R - \frac{R}{2} = \frac{R}{2} \leq FP,$$

and for the latter clearly $FP < OP$ because OP is greater then its perpendicular projection. Therefore, $CP < OP$, indeed.

Second solution. Following the lines of the first solution we shall prove the inequality $CP < OP$ once more. Since $AB = 2R \sin \gamma$ and $AC = 2R \sin \beta$,

$$BP - CP = AB \cos \beta - AC \cos \gamma =$$
$$= 2R(\sin \gamma \cos \beta - \sin \beta \cos \gamma) = 2R \sin(\gamma - \beta).$$

By the condition

$$30° \leq \gamma - \beta < \gamma < 90°, \quad \text{and thus} \quad BP - CP \geq 2R \sin 30° = R,$$

therefore $BP - CP \geq R$. Hence $R + OP = OB + OP > BP \geq R + CP$, yielding $OP > CP$.

2001/2. *Prove that*

$$(1) \qquad \frac{a}{\sqrt{a^2 + 8bc}} + \frac{b}{\sqrt{b^2 + 8ca}} + \frac{c}{\sqrt{c^2 + 8ab}} \geq 1$$

for all positive real numbers a, b, c.

First solution. Simplify the fractions on the l. h. s. of (1) by a, b and c, respectively and introduce new variables as $x = \dfrac{bc}{a^2}$, $y = \dfrac{ca}{b^2}$, $z = \dfrac{ab}{c^2}$. The claim hence becomes

$$(2) \qquad \frac{1}{\sqrt{1 + 8x}} + \frac{1}{\sqrt{1 + 8y}} + \frac{1}{\sqrt{1 + 8z}} \geq 1,$$

where

$$(3) \qquad xyz = 1.$$

With the notations $r = \sqrt{1 + 8x}$, $s = \sqrt{1 + 8y}$, $t = \sqrt{1 + 8z}$ (2) becomes

$$\frac{1}{r} + \frac{1}{s} + \frac{1}{t} \geq 1,$$

that is

$$(4) \qquad (rs + st + tr)^2 \geq (rst)^2.$$

Observe now that

$$(rs + st + tr)^2 = (rs)^2 + (st)^2 + (tr)^2 + 2rst(r + s + t) =$$
$$= (1 + 8x)(1 + 8y) + (1 + 8y)(1 + 8z) + (1 + 8z)(1 + 8x) + 2rst(r + s + t) =$$
$$= 3 + 16(x + y + z) + 64(xy + yz + zx) + 2rst(r + s + t),$$

Expanding the r. h. s. of (4) and using (3)

$$(rst)^2 = (1 + 8x)(1 + 8y)(1 + 8z) = 1 + 8(x + y + z) + 64(xy + yz + zx) + 512.$$

Substituting these results in (4), rearranging and dividing by 2 we get

$$(5) \qquad 1 + 4(x + y + z) + rst(r + s + t) \geq 256.$$

Incorporating (3) in the following *AM–GM* inequalities:

$$(6) \qquad x + y + z \geq 3\sqrt[3]{xyz} = 3;$$

$$xy + yz + zx \geq 3\sqrt[3]{x^2 y^2 z^2} = 3;$$

$$r + s + t \geq 3(rst)^{\frac{1}{3}};$$

the latter implies:

$$(7) \qquad rst(r + s + t) \geq 3(rst)^{\frac{4}{3}} = 3[(1 + 8x)(1 + 8y)(1 + 8z)]^{\frac{2}{3}} =$$
$$= 3[1 + 8(x + y + z) + 64(xy + yz + zx) + 512]^{\frac{2}{3}}.$$

Now the l. h. s. of (5) can be estimated using the results in (6) and (7).

$$1 + 4(x + y + z) + rst(r + s + t) \geq 1 + 4 \cdot 3 + 3[1 + 8 \cdot 3 + 64 \cdot 3 + 512]^{\frac{2}{3}} = 256,$$

and the proof is finished.

One can check that equality holds in the inequalities applied in the course of the solution if and only if $x = y = z = 1$. This is the case also in (2) and thus we have equality in (1) if and only if $a = b = c$.

Second solution. The fractions on the l. h. s. of (1) can be estimated separately. We show that

$$(8) \qquad \frac{a}{\sqrt{a^2 + 8bc}} \geq \frac{a^{\frac{4}{3}}}{a^{\frac{4}{3}} + b^{\frac{4}{3}} + c^{\frac{4}{3}}}.$$

This is clearly equivalent to the following inequality:

$$\left(a^{\frac{4}{3}} + b^{\frac{4}{3}} + c^{\frac{4}{3}}\right)^2 \geq a^{\frac{2}{3}}(a^2 + 8bc) = \left(a^{\frac{4}{3}}\right)^2 + 8a^{\frac{2}{3}}bc.$$

Rearranging

$$\left(a^{\frac{4}{3}} + b^{\frac{4}{3}} + c^{\frac{4}{3}}\right)^2 - \left(a^{\frac{4}{3}}\right)^2 \geq 8a^{\frac{2}{3}}bc.$$

Factorizing the l. h. s. and applying the AM–GM inequality to the arising factors

$$(a^{\frac{4}{3}} + a^{\frac{4}{3}} + b^{\frac{4}{3}} + c^{\frac{4}{3}})(b^{\frac{4}{3}} + c^{\frac{4}{3}}) \geq 4\sqrt[4]{a^{\frac{8}{3}}b^{\frac{4}{3}}c^{\frac{4}{3}}} \cdot 2\sqrt{b^{\frac{4}{3}}c^{\frac{4}{3}}} =$$

$$= 4a^{\frac{2}{3}}b^{\frac{1}{3}}c^{\frac{1}{3}} \cdot 2b^{\frac{2}{3}}c^{\frac{2}{3}} = 8a^{\frac{2}{3}}bc,$$

yields (8).

Similarly

$$\frac{b}{\sqrt{b^2 + 8ca}} \geq \frac{b^{\frac{4}{3}}}{a^{\frac{4}{3}} + b^{\frac{4}{3}} + c^{\frac{4}{3}}} \quad \text{and} \quad \frac{c}{\sqrt{c^2 + 8ab}} \geq \frac{c^{\frac{4}{3}}}{a^{\frac{4}{3}} + b^{\frac{4}{3}} + c^{\frac{4}{3}}}.$$

The sum of these three inequalities (including (8)) proves (1). Indeed

$$\frac{a}{\sqrt{a^2 + 8bc}} + \frac{b}{\sqrt{b^2 + 8ca}} + \frac{c}{\sqrt{c^2 + 8ab}} \geq \frac{a^{\frac{4}{3}} + b^{\frac{4}{3}} + c^{\frac{4}{3}}}{a^{\frac{4}{3}} + b^{\frac{4}{3}} + c^{\frac{4}{3}}} = 1.$$

Remark. A possible generalization of the problem is the following inequality: if $\lambda \geq 0$ then

$$\frac{a}{\sqrt{a^2 + \lambda bc}} + \frac{b}{\sqrt{b^2 + \lambda ca}} + \frac{c}{\sqrt{c^2 + \lambda ab}} \geq \frac{3}{\sqrt{1 + \lambda}}.$$

2001/3. *Twenty-one girls and twenty-one boys took part in a mathematical contest.*

(1) *Each contestant solved at most six problems.*

(2) *For each girl and each boy, at least one problem was solved by both of them.*

Prove that there was a problem that was solved by at least three girls and at least three boys.

Solution. Let us prepare a 21×21 array whose columns and rows indicate the boys and the girls, respectively. Write in each field the number of a problem solved by both the boy and the girl corresponding to the column and the row of the given field. By (2) one can find such a problem so there is at least one number for each field. (If there are more of them then any one can be chosen.)

Consider now the columns. By (1) there are at least six kind of numbers in each column. If some number is occuring at least three times in a column then each occurrence of this number is coloured blue in that column. We claim that thus there are at least 11 blue fields in any column. The contrary yields at least 11 uncoloured fields in that column. Since the numbers in these uncoloured fields occur at most twice, there must be at least six numbers entered in the uncoloured fields of this column. Since the boy corresponding to this column has solved at most six problems there can be no more than twelve fields containing numbers in this colum, contradicting (2).

With at least 11 blue fields in each column there are more than half of the entires are blue in our array.

Let us turn to the rows now. Proceeding as above and painting red those fields in any row whose number occurs at least three times in that row we get that more than half of the fields are red. The pigeonhole principle now implies that there must be a field that is both blue and red. Therefore, the problem whose number is contained in that field has been solved by at least three boys (because the field is blue) and also by at least three girls (because the field is red).

2001/4. *Let n be an odd integer greater than 1, and let $k_1, k_2, \ldots, k_n$ be given integers. For each of the $n!$ permutations $a = (a_1, a_2, \ldots, a_n)$ of $1, 2, \ldots, n$ let*

$$S(a) = \sum_{i=1}^{n} k_i a_i.$$

Prove that there are two permutations b and c, $b \neq c$, such that $n!$ is a divisor of $(S(b) - S(c))$.

Solution. Denote the $n!$ permutations by $A_1, A_2, \ldots, A_{n!}$, respectively. Assume that the claim is false that is $n! \nmid S(A_i) - S(A_j)$ for any $i \neq j$. There are $n!$ remainders when dividing by $n!$ and each of these remainders occurs exactly once among the sums $S(A_i)$ by our assumption. Adding these sums we get

$$\sum_{i=1}^{n!} S(A_i) \equiv 1 + 2 + \cdots + n! = \left(\frac{n!+1}{2}\right) \cdot n! \not\equiv 0 \pmod{n!}.$$

We have used that $n! + 1$ is odd and thus $\dfrac{n!+1}{2}$ is not an integer.

Let us compute the very same sum in a different way considering that each number $1, 2, \ldots, n$ occurs in exactly $(n-1)!$ permutations with coefficient k_i. Thus

$$\sum_{i=1}^{n!} S(A_i) = (n-1)! \cdot \sum_{i=1}^{n} k_i(1 + 2 + \cdots + n) =$$

$$= (n-1)! \cdot \left(\frac{n+1}{2}\right) \cdot n \cdot \sum_{i=1}^{n} k_i = n! \cdot \left(\frac{n+1}{2}\right) \cdot \sum_{i=1}^{n} k_i \equiv 0 \pmod{n!}.$$

The latter congruence clearly holds since n is odd and thus $\frac{n+1}{2}$ is a whole number and $\sum_{i=1}^{n} k_i$ is also an integer.

This is a contradiction: the same sum is both the multiple of $n!$ and not: the claim of the problem is hence true.

2001/5. *In a triangle ABC, let AP bisect $BAC\angle$, with P on BC, and let BQ bisect $ABC\angle$, with Q on CA. oldalon van. It is known that $BAC\angle = 60°$ and that $AB + BP = AQ + QB$.*

What are the possible angles of triangle ABC?

Solution. We are going to use the compulsory notations for the angles. Produce the side AB till the point P' in such a way that $BP' = BP$ and also the segment AQ till he point P'' in such a way that $QP'' = QB$. Since $AP' = AB + BP = AQ + QB = AP''$ by condition, $AP'P''$ is an equilateral triangle. (*Figure 2001/5.1.*). In the isosceles triangle $PP'B$ the base angles are equal to $\frac{\beta}{2}$ and similarly $\varphi = QP''B\angle = QBP''\angle$. The line AP is the axis of the equilateral triangle $AP'P''$ and thus $PP' = PP''$.

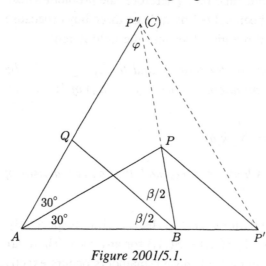

Figure 2001/5.1.

Consider now the position of the vertex C with respect to the points introduced so far. If P'' happens to be identical to C that is B, P, P'' are collinear (*Figure 2001/5.1.*), then in the isosceles triangle BQP'' we have $\varphi = \gamma = \frac{\beta}{2}$ and thus the sum of the angles in the triangle ABC is $60° + 3 \cdot \frac{\beta}{2} = 180°$. Hence

$\frac{\beta}{2} = \gamma = 40°$ and $\beta = 80°$ so the angles of the triangle ABC are $60°$, $80°$, $40°$, respectively.

Assume now that P'' is lying inside the segment QC. (*Figure 2001/5.2.* is not to scale). If this is the case then $QP''P\angle = \frac{\beta}{2}$ since P' and P'' are symmetric with respect to the line AP. Then $BP''P\angle = P''BP\angle = \frac{\beta}{2} - \varphi$ and thus the triangle BPP'' is isosceles, that is $PB = PP'' = PP' = BP'$. Therefore the triangle PBP' is equilateral and thus $\frac{\beta}{2} = 60°$, $\beta = 120°$. But this is not possible since this would imply $\alpha + \beta = 180°$. Therefore, P'' cannot lie inside the segment QC, neither on the side AC.

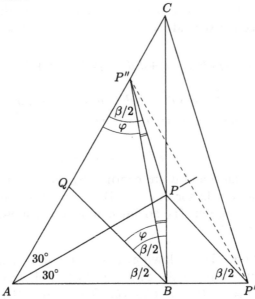

Figure 2001/5.2.

A similar contradiction shows that the point P'' cannot lie on the extension of AC over C. The only possible configuration is the one already checked: $P'' = C$, and thus the angles of triangle ABC are $60°$, $80°$ and $40°$.

Remark. The figure of the problem – like several interesting configurations of elementary geometry – can be obtained by selecting certain diagonals of a regular 18-gon. When embedded in a regular 18-gon the properties of the triangle ABC become transparent. However, the uniqueness of the configuration – as the essential part of the solution – has to be proved.

2001/6. *Let a, b, c, d be integers with $a > b > c > d > 0$. Suppose that*

(1) $ac + bd = (b + d + a - c)(b + d - a + c).$

Prove that $ab + cd$ is not a prime.

First solution. With a bit of algebra (1) becomes
$$ac + bd = [(b+d) + (a-c)][(b+d) - (a-c)] =$$
$$= (b+d)^2 - (a-c)^2 = b^2 + 2bd + d^2 - a^2 + 2ac - c^2,$$

(2)
$$b^2 + bd + d^2 = a^2 - ac + c^2.$$

Assume now, to the contrary, that $ab + cd = p$ is a prime. In what follows by congruence we mean congruence mod p. By our assumption

(3)
$$ab \equiv -cd, \qquad a^2 b^2 \equiv c^2 d^2.$$

Combining (2) and (3)
$$(b^2 + bd + d^2)b^2 = (a^2 - ac + c^2)b^2 = a^2 b^2 - ab^2 c + b^2 c^2 \equiv$$
$$\equiv c^2 d^2 + bc^2 d + b^2 c^2 = (b^2 + bd + d^2)c^2,$$
$$(b^2 - c^2)(b^2 + bd + d^2) \equiv 0.$$

Clearly $b(a - b) + c^2 > 0$ so $0 < b^2 - c^2 < ab < p$ and thus $b^2 - c^2$ is not divisible by p. Hence the second factor, $b^2 + bd + d^2$ must be its multiple. On the other hand
$$0 > b(b-a) + b(d-a) + d(d-c) = b^2 + bd + d^2 - 2ab - cd$$

implies
$$b^2 + bd + d^2 < 2ab + cd = ab + p < 2p,$$

and thus $b^2 + bd + d^2$ must be equal to p. Hence $b^2 + bd + d^2 = ab + cd$. Rearranging this equality
$$b(b + d - a) = d(c - d).$$

Observe now that b and d are coprime by $ab + cd = p$ and thus in the latter equality b must be a factor of $c - d$. However, this is not possible, because $0 < c - d < b$ by assumption. The arising contradiction shows that $ab + cd$ cannot be a prime number, indeed.

Second solution. The following solution might reveal of the possible origin of the problem. Consider the cyclic quadrilateral $ABCD$ and let its sides and angles be $AB = a$, $BC = d$, $CD = b$, $DA = c$ and $BAD\angle = 60°$, $BCD\angle = 120°$, respectively. This quadilateral clearly does exist since condition (2) yields the square of the diagonal BD when combined with the cosine rule (*Figure 2001/6.1.*):

(4)
$$BD^2 = b^2 + bd + d^2 = a^2 - ac + c^2.$$

Denote $ABC\angle$ by α. Accordingly $ADC\angle = 180° - \alpha$, and thus
$$AC^2 = a^2 + d^2 - 2ad \cos \alpha = b^2 + c^2 + 2bc \cos \alpha.$$

Isolating $2 \cos \alpha$
$$2 \cos \alpha = \frac{a^2 + d^2 - b^2 - c^2}{ad + bc}.$$

Substituting this into the first form of AC^2 we obtain
$$AC^2 = a^2 + d^2 - ad \cdot \frac{a^2 + d^2 - b^2 - c^2}{ad + bc} =$$

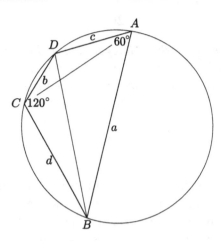

Figure 2001/6.1.

$$= \frac{a^3 d + ad^3 + a^2 bc + bcd^2 - a^3 d - ad^3 + ab^2 d + ac^2 d}{ad + bc},$$

(5)
$$AC^2 = \frac{(ab + cd)(ac + bd)}{ad + bc}.$$

By Ptolemy's theorem $AC^2 \cdot BD^2 = (ab + cd)^2$ and substituting here the results (4) and (5) we get

(6)
$$(ac + bd)(a^2 - ac + c^2) = (ab + cd)(ad + bc).$$

Observe now that the given ordering of a, b, c, d implies

(7)
$$ab + cd > ac + bd > ad + bc.$$

Indeed, if $a > b > c > d > 0$ then $(a - d)(b - c) > 0$ and $(a - b)(c - d) > 0$.

After these preliminaries let us turn to the actual proof and assume that as opposed to the claim $ab + cd$ is a prime. Since a prime number is coprime to any smaller positive integer, $ab + cd$ and $ac + bd$ are coprime by (7). Hence (6) implies that $ac + bd$ divides $ad + bc$ which is also excluded by (7). So $ab + cd$ cannot be a prime number, indeed.

Remarks. 1. In fact, the geometric environment of the second solutions was the key to identity (6). This, however, can be done by pure algebra using only (2).

$$(ac + bd)(a^2 - ac + c^2) = (ac + bd)(b^2 + bd + d^2) =$$
$$= ac(b^2 + bd + d^2) + bd(a^2 - ac + c^2) = ab^2 c + acd^2 + a^2 bd + bc^2 d =$$
$$= ab \cdot bc + cd \cdot ad + ab \cdot ad + cd \cdot bc = (ab + cd)(ad + bc).$$

2. One can actually find positive integers satisfying (1), the quadruples $(21, 18, 14, 1)$ or $(65, 50, 34, 11)$, for example.

2002.

2002/1. *Let n be a positive integer. Let T be the set of points (x, y) in the plane where x and y are non-negative integers and $x + y < n$. Each point of T is coloured red or blue. If a point (x, y) is red then so are all points (x', y') of T with both $x' \leq x$ and $y' \leq y$. Define an X-set to be a set of n blue points having distinct x-coordinates and a Y-set to be a set of n blue points having distinct y-coordinates. Prove that the number of X-sets is equal to the number of Y-sets.*

Solution. To get started let's have a closer look at the sets of the problem. T is the set of those lattice points that are on the legs and in the interior of the right triangle formed by the points $O(0,0)$, $A(n,0)$, $B(0,n)$. Since the equation of the hypotenuse is $x + y = n$, the lattice points on this line do not belong to T. As for the colouring we have to keep in mind that wherever there is a red point on the boundary of the triangle OAB, i. e. on the coordinate axes then, starting from the origin, every lattice point has to be also red on the respective axis; an interior red point R, on the other hand, forces each lattice point of the rectangle whose diagonal vertices are the origin and R and its sides are parallel to the coordinate axes also to be red. Therefore, the boundary of the red region is a kind of staircase formed by segments parallel to the coordinate axes. The points that are not red are the blue ones. (*Figure 2002/1.1.*).

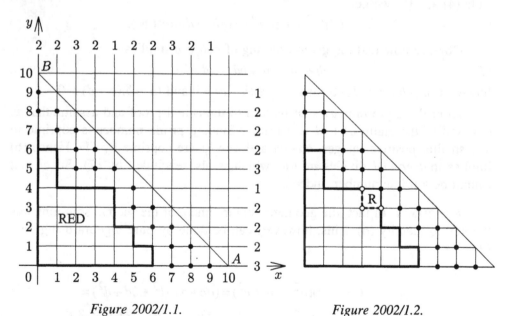

Figure 2002/1.1.   Figure 2002/1.2.

Denote the number of blue points on the lines $x = i$ and $y = i$ by a_i and b_i, respectively ($i = 0, 1, \ldots, n - 1$). (On the diagram $n = 10$ and the corresponding a_i, and b_i values are also indicated.)

Any X-set can be produced by choosing one of the a_0 blue points on the line $x = 0$ (there are a_0 ways to do that), another one from the a_1 blue points on the line $x = 1$ (this can be done in a_1 ways) and so on. An X-set hence can be constructed in exacly $a_0 a_1 \ldots a_{n-1}$ ways, this is the total number of X-sets. Similarly, there are $b_0 b_1 \ldots b_{n-1}$ Y-sets. It should be noted that if there happen to be red points on the line $x + y = n - 1$ then both products are equal to zero.

With a bit of inspection one can see that the n-tuples $a_0, a_1, \ldots, a_{n-1}$ and $b_0, b_1, \ldots, b_{n-1}$, in fact, consist of the same numbers and this seems to be the case for other legal arrangements of the red points as well. That's what we are going to prove in general; it clearly implies the claim:

(1) $$a_0 a_1 a_2 \ldots a_{n-1} = b_0 b_1 b_2 \ldots b_{n-1}.$$

We proceed by mathematical induction on the number of red points keeping the size n of the diagram fixed.

This is obvious if the number of red points is 0 or 1. We shall prove now that it holds for any number of red points once it is true if their number is one less. Choose a red point $R(x, y)$ in such a way that $x + y$ is as big as possible; differently speaking the closest one to the line $x + y = n$ is taken (*Figure 2002/1.2.*). (If there happen to be more of them then any one will do.) Let the number of blue points on the lines through R parallel to the axes be a_x and b_y, respectively. Clearly $a_x = b_y = (n - 1) - (x + y)$. Changing the colour of R to blue clearly yields a legal colouring of the lattice points and this adjustment affects a_x and b_y only: the former becomes $a_x + 1$ and the latter $b_y + 1$; by the induction hypothesis the n-tuples

$$a_0, a_1, \ldots, a_x + 1, \ldots, a_{n-1} \quad \text{and} \quad b_0, b_1, \ldots, b_y + 1, \ldots, b_{n-1}$$

are formed by the very same numbers. Since $a_x = b_y$, we also have

$$a_0, a_1, \ldots, a_x, \ldots, a_{n-1} \quad \text{and} \quad b_0, b_1, \ldots, b_y, \ldots, b_{n-1}$$

and thus (1) holds indeed.

2002/2. *Let BC be a diameter of the circle Γ with centre O. Let A be a point on Γ such that $0° < AOB\angle < 120°$. Let D be the midpoint of the arc $\overset{\frown}{AB}$ not containing C. The line through O parallel to DA meets the line AC at J. The perpendicular bisector of OA meets Γ at E and F. Prove that J is the incentre of the triangle CEF.*

Solution. Denote the radius of circle Γ by R and let the perpendicular bisector of OA meet the arc $\overset{\frown}{AC}$ at F. We start with two initial observations.

i) The endpoints O and A of the radius and the points E and F form a rhombus $OFAE$. The diagonal OA divides it into two equilateral triangles, the other diagonal EF halves the angles at its endpoints into $30°$-$30°$ (*Figure 2002/2.1.*). The length of the sides is clearly R.

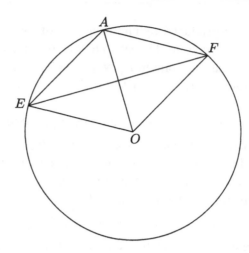

Figure 2002/2.1.

ii) In an arbitrary triangle ABC denote the incentre by K and the second intersection of the bisector of the angle C with the circumcircle by C'. Then $C'A = C'K = C'B$ and $AKB\angle = \dfrac{\alpha+\gamma}{2} + \dfrac{\beta+\gamma}{2} = \dfrac{\alpha+\beta+\gamma}{2} + \dfrac{\gamma}{2} = 90° + \dfrac{\gamma}{2}$ (*Figure 2002/2.2.*). Indeed, the central angle and the external angle theorems imply that there are equal angles on the bases AK and BK of the triangles AKC' and BKC', respectively. It follows now that the angle subtended by AB at the point K is equal to $90° + \dfrac{\gamma}{2}$ and there is exactly one point on the segment CC' having this property.

Accordingly, in order to demonstrate that a certain point K on the segment CC' is, in fact, the incentre of the triangle ABC it is enough to verify any one of the following propositions:

  1. K is incident to the bisector of the angle A; or

  2. its distance from C' is equal to $C'A = C'B$; or

  3. it is an interior point of the triangle at which the angle subtended by the side AB is equal to $90° + \dfrac{\gamma}{2}$.

Let us turn to the solution of the problem (*Figure 2002/2.3.*). Since the arcs $\overset{\frown}{BD}$ and $\overset{\frown}{DA}$ are equal, OD is bisecting the angle $AOB\angle$ that is the external angle of the isosceles triangle AOC. Hence OD is parallel to the base AC of this triangle. Since DA and OJ are also parallel, the quadrilateral $ADOJ$ is a parallelogram and thus $AJ = OD = R$. On the other hand halving the arc $\overset{\frown}{EF}$ the line AC bisects $ECF\angle$. Observe, finally, that $ECF\angle = 60°$ because the corresponding cenral angle is equal to $120°$.

The condition on $AOB\angle = \alpha$ assures that the point J is, in fact, lying inside the triangle CEF as it is shown on the diagram. Indeed, the vectors $\overrightarrow{OJ}$

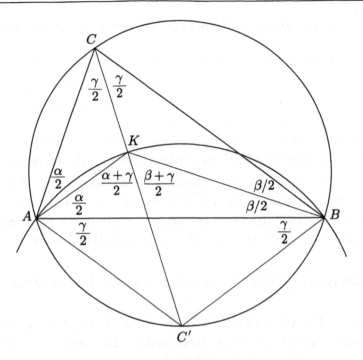

Figure 2002/2.2.

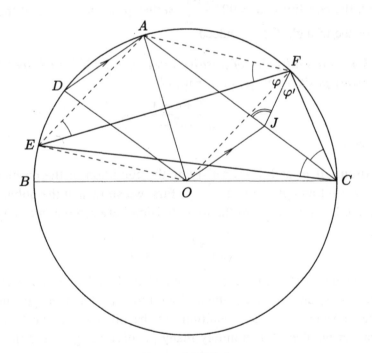

Figure 2002/2.3.

and $\overrightarrow{DA}$ are parallel and thus the vector $\overrightarrow{OC}$ should be rotated counterclockwise by $90° - \dfrac{3\alpha}{4}$ to be parallel to $\overrightarrow{OJ}$ because the vector $\overrightarrow{DA}$ makes positive angle

$90° - \dfrac{3\alpha}{4}$ with the direction of the vector $\overrightarrow{OC}$. Accordingly, if $\alpha < 120°$ then the halfline OJ and the point A are on the same sides of the line BC and thus OJ cuts AC inside the triangle CEF. Exactly this restriction on α is the condition for this to happen. Now we show that J is the incentre of the triangle CEF; we provide three different arguments corresponding to propositions 1. 2. and 3. above. Angles of magnitude $30°$ are marked with a single arc on the diagram.

1. Since $AF = AJ = R$, the triangle AFJ is isosceles. If $EFJ\angle = \varphi$ and $CFJ\angle = \varphi'$ then the base angles of AFJ are equal to $\varphi + 30°$. As the complementary angle to $FJC\angle$ the very same angle is equal to $\varphi' + 30°$. Since $\varphi + 30° = \varphi' + 30°$ and thus $\varphi = \varphi'$, the point J is incident to the bisector of the angle F, indeed.

2. Clearly $AJ = AF = AE = R$ and proposition 2. immediately yields the claim.

3. Since $AE = AO = AJ = AF = R$, the points E, O, J, F are on a circle of centre A and radius R. The angle subtended by the chord EF of this circle at the point O is equal to $120°$. Therefore the same angle is subtended by EF at J. Finally, the equality $120° = 90° + \dfrac{60°}{2}$ settles proposition 3. and thus J is the incentre of the triangle CEF, indeed.

2002/3. *Find all pairs of positive integers $m, n \geq 3$ such that there exist infinitely many positive integers a for which*

$$\frac{a^m + a - 1}{a^n + a^2 - 1}$$

is an integer.

Solution. If m and n are satisfying the condition of the problem then let $f(x) = x^m + x - 1$ and $g(x) = x^n + x^2 - 1$. First we show that the polynomial $g(x)$ divides $f(x)$. Let's write down the long division between the two polynomials:

(1) $$\frac{f(x)}{g(x)} = h(x) + \frac{r(x)}{g(x)}.$$

The leading coefficient of $g(x)$ is equal to 1 and thus the coefficients of the quotient $h(x)$ are whole numbers. Therefore, it admits integer numbers for every integer value of x. By condition, on the other hand, the l. h. s. of (1) admits integer numbers for infinitely many positive integers and thus the same holds for $\dfrac{r(x)}{g(x)}$. Since $\deg r < \deg g$ the ratio $\dfrac{r(x)}{g(x)}$ of the long division tends to zero if x tends to infinity. If a sequence of whole numbers tends to zero then the sequence must admit the value zero infinitely many times, too: the ratio $\dfrac{r(x)}{g(x)}$ has infinitely many roots and so does the numerator $r(x)$. Therefore, it is identically zero. Since it is the remainder of the division, $g(x) \mid f(x)$, indeed. Note that this relation also implies $m \geq n$.

Consider now the following multiple of $g(x)$:

$$(x+1)f(x) - g(x) = x^n(x^{m-n+1} + x^{m-n} - 1).$$

Denote now $m - n \geq 0$ by k. Since x^n and $g(x) = x^n + x^2 - 1$ are coprime we clearly have

(2) $$g(x) = x^n + x^2 - 1 \mid x^{k+1} + x^k - 1.$$

For the final blow we are going to make use of the continuity of the polynomial function $g(x)$. Since $g(0) < 0 < g(1)$ the function $g(x)$ certainly has a root between 0 and 1, there is some $0 < \alpha < 1$ such that $g(\alpha) = 0$. The same α is a root of the polynomial $x^{k+1} + x^k - 1$ by (2). Therefore

(3) $$\alpha^n + \alpha^2 = \alpha^{k+1} + \alpha^k = 1.$$

Now $n \geq 3$ by condition and $n \leq k+1$ by (2). Hence

$$k \geq n - 1 \geq 2.$$

Since $0 < \alpha < 1$ this implies $\alpha^n \geq \alpha^{k+1}$ and $\alpha^2 \geq \alpha^k$. But now (3) restricts the inequalities to $\alpha^n = \alpha^{k+1}$ and $\alpha^2 = \alpha^k$, that is $n = k+1$ and $2 = k$ and thus $m = 5$ and $n = 3$.

Since $a^5 + a - 1 = (a^3 + a^2 - 1)(a^2 - a + 1)$ the conditions hold for $m = 5$ and $n = 3$. Therefore, the only solution of the problem is the pair $(5; 3)$.

2002/4. *Let n be an integer greater than 1. The positive divisors of n are $d_1, d_2, \ldots, d_k$ where*

$$1 = d_1 < d_2 < \ldots < d_k = n.$$

Define $D = d_1 d_2 + d_2 d_3 + \ldots + d_{k-1} d_k$.

(a) *Prove that $D < n^2$.*

(b) *Determine all n for which D is a divisor of n^2.*

Solution. If d is a divisor of n then so is $\dfrac{n}{d}$ because $\dfrac{n}{d} \cdot d = n$. The terms of the pair d and $\dfrac{n}{d}$ are called complementary divisors with respect to n. (It may happen that d is equal to its pair, namely if $n = p^2$ then $p = \dfrac{n}{p}$). The set of the complentary pairs of n's divisors is clearly identical to the set of the divisors; their ordering, however, is reversed

$$1 = \frac{n}{d_k} < \frac{n}{d_{k-1}} < \ldots < \frac{n}{d_2} < \frac{n}{d_1} = n.$$

Replace now the divisors in D by their respective complementary pairs:

$$D = d_1 d_2 + d_2 d_3 + \ldots + d_{k-1} d_k = \frac{n}{d_k} \cdot \frac{n}{d_{k-1}} + \frac{n}{d_{k-1}} \cdot \frac{n}{d_{k-2}} + \ldots + \frac{n}{d_2} \cdot \frac{n}{d_1} =$$

$$= n^2 \left(\frac{1}{d_1 d_2} + \frac{1}{d_2 d_3} + \ldots + \frac{1}{d_{k-1} d_k} \right) \leq$$

$$\leq n^2 \left(\frac{d_2 - d_1}{d_1 d_2} + \frac{d_3 - d_2}{d_2 d_3} + \ldots + \frac{d_k - d_{k-1}}{d_{k-1} d_k} \right) =$$

$$= n^2 \left(\frac{1}{d_1} - \frac{1}{d_2} + \frac{1}{d_2} - \frac{1}{d_3} + \ldots + \frac{1}{d_{k-1}} - \frac{1}{d_k} \right) = n^2 \left(\frac{1}{d_1} - \frac{1}{d_k} \right) < \frac{n^2}{d_1} = n^2,$$

establishing the first part of the claim.

Turning to $b)$ note first that for n prime D divides n^2 since now $1 = d_1 < < d_2 = p = n$ and $D = d_1 d_2 = p = n$. For n composite, on the other hand, clearly $k \geq 3$. The smallest nontrivial divisor of n is some prime number p being that of n^2 as well; accordingly

$$1 = d_1 < d_2 = p < \ldots < d_{k-1} = \frac{n}{p} < d_k = n.$$

Hence $D = d_1 d_2 + \ldots + d_{k-1} d_k > d_{k-1} d_k = n \cdot \frac{n}{p} = \frac{n^2}{p}$, yielding $p > \frac{n^2}{D}$.

Assume now that D divides n^2. Its pair $\frac{n^2}{D}$ is also a divisor and by the first part's estimation $\frac{n^2}{D} > 1$. Therefore,

$$1 < \frac{n^2}{D} < p$$

contradicting to the choice of p as the smallest nontrivial divisor of n^2.

As a result D divides n^2 if and only if n is a prime number.

2002/5. *Find all functions f from the set $\mathbf{R}$ of real numbers to itself such that*

(1) $$(f(x) + f(z))(f(y) + f(t)) = f(xy - zt) + f(xt + yz)$$

for all x, y, z, t in $\mathbf{R}$.

Solution. The flow of the solution is exemplary: if there are functional equations around one usually proceeds by deducing the consequences of particular substitutions.

Putting $y = z = t = 0$ first (1) becomes

$$(f(x) + f(0))(f(0) + f(0)) = f(0) + f(0),$$

(2) $$f(0)(f(x)+f(0))=f(0).$$

Setting x also equal to zero (2) yields
$$2f^2(0)=f(0), \qquad f(0)(2f(0)-1)=0$$
and thus there are two possible values for $f(0)$: it is either $\dfrac{1}{2}$ or 0.

   $i)$ If $f(0)=\dfrac{1}{2}$ then (2) implies $f(x)+\dfrac{1}{2}=1$ that is
$$f(x)=\frac{1}{2}$$
for every real number x. It clearly satisfies (1):
$$\left(\frac{1}{2}+\frac{1}{2}\right)\left(\frac{1}{2}+\frac{1}{2}\right)=\frac{1}{2}+\frac{1}{2}.$$

   $ii)$ Consider now the other possibility, $f(0)=0$. Plug $z=t=0$ in (1) once more. The result is

(3) $$f(x)f(y)=f(xy),$$
our function f is hence *multiplicative*. Let $x=y=1$ in (3): $f^2(1)=f(1)$: either
$$\text{A: } f(1)=0, \quad \text{or} \quad \text{B: } f(1)=1.$$

A In this case we have $f(0)=0$, $f(1)=0$. Plugging $y=1$ in (3) implies
$$f(x)=0$$
for every real number x; this is again a solution, since $(0+0)(0+0)=0+0$.

   B. We are left with the case $f(0)=0$, $f(1)=1$. Plug now $x=0$, $y=t=1$ in (1):
$$(f(0)+f(z))(f(1)+f(1))=f(-z)+f(z),$$
that is
$$2f(z)=f(-z)+f(z),$$
$$f(z)=f(-z),$$
so our function f is even. If $y=x$ in (3) then

(4) $$f(x^2)=f^2(x)\geq 0.$$
Therefore, f admits non negative values for $x\geq 0$ and being even $f(x)\geq 0$ for every real number x.

   Set now $x=t$?s $y=z$ in (1).
$$(f(x)+f(y))^2=f(x^2+y^2),$$
yielding
$$f(x^2+y^2)=f^2(x)+f^2(y)+2f(x)f(y)\geq f^2(x)$$
so by (4)
$$f(x^2+y^2)\geq f(x^2).$$
Therefore, f is increasing for x positive.

Setting finally $y = z = t = 1$ in (1) implies

(5) $2(f(x) + 1) = f(x-1) + f(x+1).$

Plugging $x = 2$, and 3 yields $f(2) = 4$, $f(3) = 9$. Together with the already established $f(1) = 1$ and $f(0) = 0$ it is hard to resist the conjecture $f(n) = n^2$ for n non negative. The proof is an immediate coasequence of (5) by induction: for $n = 0$ we have it; let $n > 0$ and assume that it holds for $n - 1$. If $x = n - 1$ in (5) then

$$2(f(n-1) + 1) = f(n-2) + f(n),$$
$$2((n-1)^2 + 1) = (n-2)^2 + f(n)$$
$$f(n) = 2n^2 - 4n + 2 + 2 - n^2 + 4n - 4 = n^2.$$

So $f(x) = x^2$ for every non negative whole number, indeed, and since our function is even, this also holds for any integral value of x.

Let now $x = \dfrac{p}{q}$ be an arbitrary rational number (p and q are integers.). By (3) we get

$$f\left(\frac{p}{q}\right) f(q^2) = f(pq), \quad \text{that is} \quad f\left(\frac{p}{q}\right) q^4 = p^2 q^2,$$

$$f\left(\frac{p}{q}\right) = \frac{p^2}{q^2},$$

therefore, $f(x) = x^2$ for every rational number x. To complete the proof we show now that this must hold for any real number x. It is clearly enough to prove this for positive numbers since our function is even. Assume, to the contrary, that for some $x > 0$ $f(x) < x^2$. Choose now a rational number r between $\sqrt{f(x)}$ and x:

(6) $\sqrt{f(x)} < r < x$ that is $f(x) < r^2 < x^2.$

Since $f(r) = r^2$ and f is increasing, $f(r) = r^2 \le f(x)$ contradicting to (6). Thus $f(x) < x^2$ is impossible; a similar contradiction can be deduced if $f(x) > x^2$ is assumed.

Accordingly, in the last case there is only one possibility, indeed: $f(x) = x^2$. It is, in fact, a solution since now (1) becomes the well known

$$(x^2 + z^2)(y^2 + t^2) = (xy - zt)^2 + (xt + yz)^2,$$

the two dimensional form of the so called Lagrange-identity. Reviewing the solutions, there are the following functions satisfying (1):

$$f(x) = \frac{1}{2}, \quad f(x) = 0, \quad f(x) = x^2.$$

Remark. The background of the problem is clearly the celebrated identity of Lagrange. For those familiar with this widely used result it must have been clear that the principal solution of the functional equation (1) is $f(x) = x^2$. In its simplest form as in the problem the identity is equivalent to the trigonometric form of Pithagoras: $\sin^2 \alpha + \cos^2 \alpha = 1$. Indeed, if the vectros **a** and **b** are $\mathbf{a}(x, z)$, $\mathbf{b}(t, y)$ and their angle is α then

$$\cos^2 \alpha = \frac{(xt + zy)^2}{(x^2 + z^2)(y^2 + t^2)}, \qquad \sin^2 \alpha = \frac{(xy - zt)^2}{(x^2 + z^2)(y^2 + t^2)},$$

and this can be naturally extended to higher dimensions as well. A nontrivial corollary of the identity is the inequality of Cauchy. [22]

2002/6. *Let $\Gamma_1, \Gamma_2, \ldots, \Gamma_n$ be circles of radius 1 in the plane, where $n \geq 3$. Denote their centres by $O_1, O_2, \ldots, O_n$, respectively. Suppose that no line meets more than two of the circles. Prove that*

$$\sum_{1 \leq i < j \leq n} \frac{1}{O_i O_j} \leq \frac{(n-1)\pi}{4}.$$

First solution. Note first that if the chords PQ and ST in the circle of radius R meet at the point X and $SXP\angle = 2\alpha$ then

(1) $$\widehat{PS} + \widehat{QT} = 4R\alpha$$

(Figure 2002/6.1.). ($\widehat{PS}$ denotes the length of the arc PS; angles are in circular measure.) Denote, for the proof, the inscribed angles subtended by the arcs $\widehat{PS}$ and $\widehat{QT}$ by ω and φ, respectively. Accordingly, the corresponding central angles are 2ω and 2φ and thus

$$\widehat{PS} = R \cdot 2\omega, \qquad \widehat{QT} = R \cdot 2\varphi.$$

Since 2α is external angle in the triangle PXT we have $2\alpha = \omega + \varphi$ and hence

$$\widehat{PS} + \widehat{QT} = 2R(\omega + \varphi) = 4R\alpha.$$

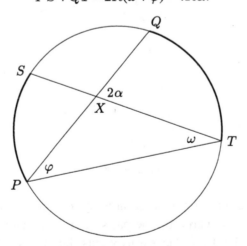

Figure 2002/6.1.

Let us cover now the given unit circles by a circle C of radius R. Consider two of the circles: their centres are O_i and O_j, respectively, their internal tangents

PQ and ST meet at X making angle 2α. These cirles are disjoint, otherwise a secant through a common point to a third circle would meet at least three of them, contradicting to the condition. Reviewing the *diagram 2002/6.2.* one can gather

$$O_iO_j = \frac{2}{\sin\alpha}.$$

Since α is acute, we clearly have

$$\alpha \geq \sin\alpha = \frac{2}{O_iO_j},$$

and combining this with (1) we get

(2)
$$\widehat{PS} + \widehat{QT} = 4R\alpha \geq \frac{8R}{O_iO_j},$$

$$\frac{1}{O_iO_j} \leq \frac{\widehat{PS} + \widehat{QT}}{8R}.$$

Denote the sum $\widehat{PS} + \widehat{QT}$ of the arcs corresponding to the centres O_i, O_j and

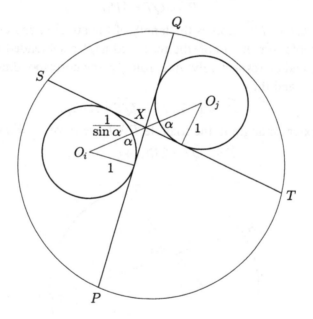

Figure 2002/6.2.

also, for simplicity, the pair of these arcs by s_{ij}. Let's try to estimate to what extent is the circle C covered by the pairs of arcs s_{ij}. Note first that any secant intersecting the circles Γ_i, Γ_j necessarily cuts the arcs s_{ij}. Consider now an arbitrary point E on C; we are going to reckon the number of arc-pairs covering this point. Draw, for this purpose, the halfline e touching C at E and begin rotating it as indicated on the *Figure 2002/6.3.* Let the first pair of circles cut by the revolving line be (Γ_1, Γ_2); E is hence incident to s_{12}. Revolving further on e leaves (say) Γ_1 and either cuts another circle, say Γ_3 while still intersecting Γ_2 or it leaves the second circle without meeting a third one. Going on it is clear that the most the rotating line can do is to intersect the consecutive pairs of circles

$$(\Gamma_1, \Gamma_2), (\Gamma_2, \Gamma_3), \ldots, (\Gamma_{n-1}, \Gamma_n)$$

and thus E is covered by the pairs of arcs $s_{12}, s_{23}, \ldots, s_{n-1,n}$, at most $n-1$ times altogether. This is clearly true for any point on C and thus the sum of the covering arcs is at most $n-1$ times the perimeter of C, that is $2R\pi(n-1)$. Since (2) can also be written as $\dfrac{1}{O_iO_j} \le \dfrac{s_{ij}}{8R}$, this estimation yields the claim. Indeed,

$$\sum_{1 \le i < j \le n} \frac{1}{O_iO_j} \le \sum_{1 \le i < j \le n} \frac{s_{ij}}{8R} \le \frac{2R\pi(n-1)}{8R} = \frac{(n-1)\pi}{4}.$$

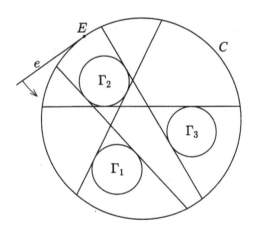

Figure 2002/6.3.

Second solution. Select any one of the given circles and denote its centre by O_i. The tangents from O_i to the given circles can be certainly drawn, since the circles are disjoint. Denote the angle of the pair of tangents to the circle of centre O_j by $2\alpha_{ij}$. Then

(1)
$$\frac{1}{O_iO_j} = \sin \alpha_{ij} \le \alpha_{ij},$$

since α_{ij} is acute.

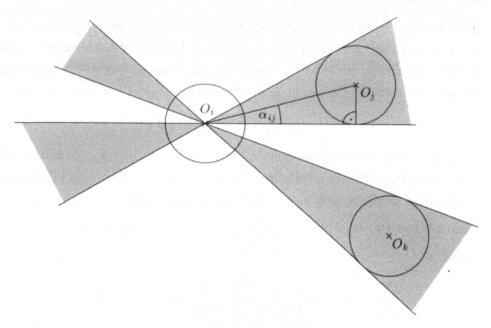

Figure 2002/6.4.

The vertically opposite angular regions formed by the respective pairs **are** shaded on the ***diagram** 2002/6.4*. Apart from O_i these regions **have no common** points by condition. Therefore, the total circular measure **of these** angles **does** not exceed 2π and we have the following estimation:

$$4 \sum_{j=1,\ j\neq i}^{n} \frac{1}{O_i O_j} = 4 \sum_{j=1,\ j\neq i}^{n} \sin\alpha_{ij} \leq 4 \sum_{j=1,\ j\neq i}^{n} \alpha_{ij} \leq 2\pi,$$

that is

$$(2) \qquad \sum_{j=1,\ j\neq i}^{n} \frac{1}{O_i O_j} \leq \frac{\pi}{2}.$$

When summing inequalities (2) for $i = 1, 2, \ldots n$ each pair $O_x O_y$ occurs twice; hence

$$(3) \qquad \sum_{1 \leq i < j \leq n} \frac{1}{O_i O_j} \leq \frac{n\pi}{4}.$$

The claim itself is slightly stronger than (3). Let's have **a** closer look at a possible improvement of our estimation. Consider the **convex hull of the** centres, choose one of its vertices as O_l and denote the external angle at O_l by φ.

Lemma: Setting O_l as O_i in the previous argument the measure of the region about O_l covered by the tangent pairs drawn to the remaining circles is at most $(2\pi - \varphi)$.

This implies the claim already because the sum of the external angles is 2π, and thus when summing the inequalities (2) we obtain the refined estimation as follows:

$$\sum_{1\leq i<j\leq n} \frac{1}{O_i O_j} \leq \frac{n\pi}{4} - \frac{1}{8} \sum_{\substack{\text{k?ls?}\\ \text{sz?gek}}} \varphi = \frac{n\pi}{4} - \frac{\pi}{4} = \frac{(n-1)\pi}{4}.$$

Proof of the Lemma. Denote, along the convex hull, the neighbouring vertices to O_l by O_k and O_m, respectively. Let the line $O_k O_m$ be e, and denote by f the line through O_l parallel to e. The width of the band bounded by these parallel lines is greater than 2, otherwise the bimedian parallel to e of the triangle $O_k O_l O_m$ would meet three of the given circles. Rotate now f about O_l, first to touch the circle about O_k at K then the circle about O_m at M. (*Figure 2002/6.5.*)

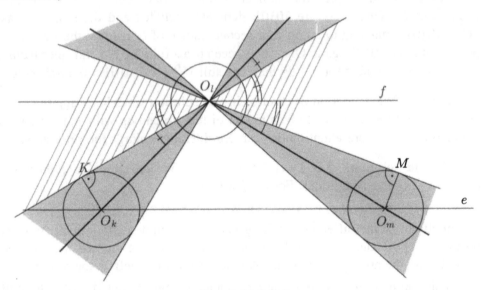

Figure 2002/6.5.

Since $MO_m = 1$, the perpendicular distances of M from the lines $O_l O_m$ and e are both less than 1. Hence, by the previous observation its distance from f is greater than 1. Accordingly, the angle $MO_l O_m \angle$ is less than the half of the angle made by the lines f and $O_l O_m$. We get a similar estimation at the other circle and thus

$$\sphericalangle \leq \sphericalangle ; \qquad \sphericalangle \leq \sphericalangle .$$

The region striped on the diagram cannot be covered by the remaining circles and the corresponding tangent pairs. Its magnitude is

$$2 \sphericalangle + 2 \sphericalangle \geq \sphericalangle + \sphericalangle + \sphericalangle + \sphericalangle = \varphi, \qquad \text{indeed.}$$

The proof of the lemma and hence the solution is complete.

2003.

2003/1. *Let A be a subset of the set $S = \{1, 2, \ldots, 1000000\}$ containing exactly 101 elements. Prove that there exist numbers t_1, t_2, $\ldots$, t_{100} in S such a way that the sets*

$$A_j = \{x + t_j \mid x \in A\} \qquad j = 1, 2, \ldots, 100$$

are pairwise disjoint.

First solution. Sets A_i and A_j share an element if and only if $x_k + t_i = x_l + t_j$ holds for some k and l. Since there are 101 possible values for k and one less for l there can be no more than 10100 such values for t_j if t_i is fixed. The demand that A_i and A_j be disjoint means that t_j has to be different from these at most 10100 numbers and from t_i as well.

We are going to pick the numbers t one by one. Let t_1 be an arbitrary element of S. Thus there are 10101 elements excluded and there are at least $10^6 - 10101$ remaining to go on. Our second choice of t_2 reduces the size of the pool by at most 10101 again. One can proceed peacefully till the end since having chosen the kth number for $k \leq 99$ there are still $10^6 - k \cdot 10101 > 0$ numbers left.

Remark. We were kind of careless in this solution when selecting the numbers one at a time; it seems to be a matter of luck to end up with exactly 100 of them. With a bit more attention one can considerably enlarge the selected set.

Second solution. Let $1 = t_1 < t_2 < t_3 < \ldots$ We are going to capture our numbers t in increasing order. For $i < j$ we have

$$x_k + t_1 \neq x_l + t_2, \qquad \text{that is} \qquad x_k - x_l \neq t_j - t_i > 0$$

by condition, so the differences arising inside A exclude certain numbers to be taken. Set $t_1 = 1$ and go on in the greedy manner: having crossed the numbers excluded by the arising differences take the smallest one left as the next t.

This can be followed below on a reasonably small A. Let $A = \{2; 7; 8\}$ with the arising differences 1, 5 and 6. In the table below the selected numbers are indicated by ✓, those excluded by 0.

S:	1	2	3	4	5	6	7	8	9	10	11	12	13	14	15	16	17	18	19	20	...
t_1	✓	0				0	0														
t_2			✓	0				0	0												
t_3					✓	0				0	0										
t_4												✓	0				0	0			
t_5														✓	0				0	0	

$\vdots$

Any number can be excluded more than once, of course, like 6 above.

Denote the number of differences arising inside A by m. Let $k \geq 0$ and assume that the numbers $t_1, t_2, \ldots, t_k$ have been already selected. What can be said about t_{k+1}? There are k numbers chosen (those marked by ✓) and at most $k \cdot m$ excluded (marked by 0). Therefore,

$$t_{k+1} \leq k + k \cdot m + 1.$$

In the problem $m \leq \binom{101}{2}$ and thus one can keep choosing the numbers t as long as

$$k + k \cdot \binom{101}{2} + 1 \leq 10^6, \qquad \text{that is} \qquad k \leq 197{,}98.$$

Thus one can find at least 198 numbers this way, a serious improvement.

Remark. In the language of graph theory the problem is closely related to Turán's theorem; the vertices of the graph are the elements of S and two of them are connected with an edge if the difference of the corresponding numbers does not appear as a difference in the set A. The task is to find a big complete subgraph [45].

2003/2. *Determine all pairs of positive integers (a, b) such that*

$$\frac{a^2}{2ab^2 - b^3 + 1}$$

is a positive integer.

Solution. For a solution a and b denote the value of the fraction by k. Since $k > 0$ and $a^2 > 0$, the denominator is also positive, $2ab^2 - b^3 + 1 > 0$. Therefore,

(1) $$a > \frac{b}{2} - \frac{1}{2b^2}, \qquad \text{that is} \qquad a \geq \frac{b}{2}.$$

Since $k \leq 1$, we also have $a^2 \leq b^2(2a - b) + 1$, or $a^2 > b^2(2a - b) \leq 0$. Accordingly, there are two possibilities. Either

(2) $$a > b \qquad \text{or} \qquad 2a = b.$$

Denote the roots of the quadratic

$$x^2 - 2kb^2 x + k(b^3 - 1) = 0$$

by a_1 and a_2, respectively. Plugging a for x yields the condition of the problem rearranged. Therefore, one of the roots is a. Since it is a whole number and so is $a_1 + a_2 = 2kb^2$, both roots are integer numbers. One can assume, by symmetry, that $a_1 \geq a_2$. Comparing this to the sum of the roots one gets $a_1 \geq k \cdot b^2 > 0$. On the other hand $a_1 \cdot a_2 = k(b^3 - 1)$, and thus

$$0 \leq a_2 = \frac{k(b^3 - 1)}{a_1} \leq \frac{k(b^3 - 1)}{kb^2} < b.$$

Combining this with (2) yields either $a_2 = 0$ or $2a_2 = b$.

If $a_2 = 0$ then $b^3 - 1 = 0$, $b = 1$, $a_1 = 2k$.

If $2a_2 = b$ (then b is even), then

$$a_1 + \frac{b}{2} = 2kb^2 \quad \text{and} \quad a_1 \cdot \frac{b}{2} = k(b^3 - 1), \quad \text{and thus} \quad k = \frac{b^2}{4}, \quad a_1 = \frac{b^4}{2} - \frac{b}{2}.$$

Summarizing our results:

$$(2l, 1) \qquad \text{or} \qquad (l, 2l) \qquad \text{or} \qquad (8l^4 - l, 2l)$$

for some positive integer l. It is obvious that these pairs are all solutions, indeed.

2003/3. *A convex hexagon is given in which any two opposite sides have the following property: the distance between their midpoints is $\sqrt{3}/2$ times the sum of their lengths. Prove that all the angles of the hexagon are equal.*

(A convex hexagon $ABCDEF$ has three pairs of opposite sides: AB and DE, BC and EF, CD and FA.)

Solution.

Lemma. Assume in the triangle PQR that the angle $QPR\angle \geq 60°$ and denote the midpoint of QR by T. Then $TP \leq \frac{\sqrt{3}}{2}QR$ and equality holds if and only if PQR is equilateral.

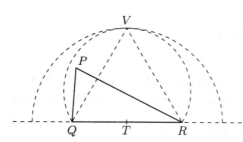

Figure 2003/3.1.

Consider the equilateral triangle VQR for which the points V and P are on the same side of the line QR. Then P is lying on the region bounded by the circumcircle of VQR. This circle is covered by the circle of centre T and radius $\frac{\sqrt{3}}{2}QR$ and the only common point is V. The lemma is hence proved.

Consider, among the diagonals AD, BE and CF the pair whose angle is the biggest. Assume that these are AD and BE, denote their common point by P, and the midpoints of AB and DE by M and N, respectively. Since their angle at P is at least 60°, the lemma applies for the triangles BPA and DPE. Therefore

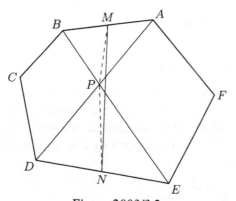

Figure 2003/3.2.

$$MN = \frac{\sqrt{3}}{2}(AB + DE) \geq$$
$$\geq PM + PN \geq MN.$$

There is equality in our chain and this can happen only if both triangles BPA and DPE are equilateral. The same reasoning yields that the "main" diagonal CF also makes $60°$ with the diagonals AD and BE, the same as the angle of any "main" diagonal with the sides starting from its vertices. Accordingly, all the angles of our hexagon are equal to $120°$.

2003/4. *Let $ABCD$ a cyclic quadrilateral. Let P, Q and R be the feet of the perpendiculars from D to the lines BC, CA and AB respectively. Show that $PQ = QR$ if and only if the bisectors of $ABC\angle$ and $ADC\angle$ meet on AC.*

Solution. The assumption that the quadrilateral be cyclic is not necessary; in the following solution we shall not use this condition.

Denote, in the triangle ABC the angles at A and C by α and γ, respectively. P and Q are incident to the Thales circle of CD and $\sin PCQ\angle = \sin\gamma$ or $PQ = CD \cdot \sin\gamma$. Similarly $QR = AD \cdot \sin\alpha$. By condition

$$PQ = CD \cdot \sin\gamma = AD \cdot \sin\alpha = QR.$$

Rearranging and applying the sine rule for the triangle ABC yields

$$\frac{CD}{AD} = \frac{\sin\alpha}{\sin\gamma} = \frac{CB}{AB}.$$

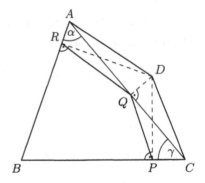

Figure 2003/4.1.

By the angle bisector theorem this holds if and only if the bisectors of the angles $ABC\angle$ and $ADC\angle$ meet the segment AC at the very same point.

Remark. If D is in fact incident to the circumcircle of the triangle ABC then P, Q and R are collinear; they are lying on the so called Simson line of the triangle.

2003/5. *Let n be a positive integer and $x_1, x_2, \ldots, x_n$ be real numbers with $x_1 \le x_2 \le \ldots \le x_n$.*

(a) *Prove that*

$$\left(\sum_{i=1}^{n}\sum_{j=1}^{n}|x_i - x_j|\right)^2 \le \frac{2(n^2-1)}{3}\sum_{i=1}^{n}\sum_{j=1}^{n}(x_i - x_j)^2.$$

(b) *Show that equality holds if and only if $x_1, \ldots, x_n$ is an arithmetic sequence.*

Solution. Increasing or decreasing each x_i by the same value does not affect either of the claims so

$$\sum_{i=1}^{n} x_i = 0$$

can be assumed. Then

$$\sum_{i=1}^{n}\sum_{j=1}^{n}|x_i - x_j| = 2\sum_{i>j}(x_i - x_j) = 2\sum_{i=1}^{n}(2i - n - 1)x_i.$$

Applying Cauchy's inequality for the l. h. s. yields

(1) $\quad\left(\sum_{i=1}^{n}\sum_{j=1}^{n}|x_i - x_j|\right)^2 \le 4\sum_{i=1}^{n}(2i - n - 1)^2 \sum_{i=1}^{n}x_i^2 = 4\frac{n(n+1)(n-1)}{3}\sum_{i=1}^{n}x_i^2.$

The last equality can be obtained by the well known formula for the sum of the square numbers. Indeed

$$\sum_{i=1}^{n}(2i - n - 1)^2 = \sum_{i=1}^{n}4i^2 - 2\sum_{i=1}^{n}2i(n+1) + n(n+1)^2 =$$

$$= 4\frac{n(n+1)(2n+1)}{6} - 2n(n+1)^2 + n(n+1)^2 = \frac{n(n+1)(n-1)}{3}.$$

Let's turn now to the right hand side.

(2) $\quad\sum_{i=1}^{n}\sum_{j=1}^{n}(x_i - x_j)^2 = n\sum_{i=1}^{n}x_i^2 - 2\sum_{i=1}^{n}x_i\sum_{j=1}^{n}x_j + n\sum_{j=1}^{n}x_j^2 = 2n\sum_{i=1}^{n}x_i^2.$

Combining (1) and (2) our inequality now follows.

$$\left(\sum_{i=1}^{n}\sum_{j=1}^{n}|x_i - x_j|\right)^2 \le \frac{2(n+1)(n-1)}{3} \cdot 2n\sum_{i=1}^{n}x_i^2 = \frac{2(n^2-1)}{3}\sum_{i=1}^{n}\sum_{j=1}^{n}(x_i - x_j)^2.$$

Since the integers $(2i - n - 1)$ form an arithmetic sequence in (1), there is equality in Cauchy's inequality if and only if the same holds for the numbers $x_1, x_2, \ldots, x_n$.

2003/6. *Let p be a prime number. Prove that there exists a prime number q such that for any integer n, the number $n^p - p$ is not divisible by q.*

Solution. Since

$$S = \frac{p^p - 1}{p - 1} = 1 + p + p^2 + \ldots + p^{p-1} \equiv 1 + p \pmod{p^2},$$

the sum S has some prime divisor q whose remainder when divided by p^2 is not 1. We prove that this prime divisor can be chosen as q.

Assume that $n^p \equiv p \pmod{q}$ for some integer n. Then, by q's choice

$$(n^p)^p = n^{p^2} \equiv p^p \equiv 1 \pmod{q}.$$

On the other hand $n^{q-1} \equiv 1 \pmod{q}$ by the little Fermat theorem [32]. Consider the indices of n occurring so far. Since $p^2 \nmid q-1$, and thus $(p^2, q-1) \mid p$ yielding $n^p \equiv 1 \pmod{q}$.

Accordingly, $n^p \equiv p \pmod{q}$ by assumption and $n^p \equiv 1 \pmod{q}$ by the choice of q. Hence $p \equiv 1 \pmod{q}$. But then

$$S = 1 + p + p^2 + \ldots + p^{p-1} \equiv p \pmod{q}.$$

Since $q \mid S$, the last relation implies $p \equiv 0 \pmod{q}$, contradicting to our assumption. Hence $n^p - p$ is not a multiple of q, indeed.

Remark. Even if the solution is reasonably compact, the actual identification of the prime number q – this is silently done in the first paragraph without any particular fuss – makes the problem extremely difficult. Just for your information: out of the total 457 participants of this Olympiad there were 24 who solved problem No. 3 (!) and 26 who principally solved problem No. 6.

2004.

2004/1. *Let ABC be an acute-angled triangle with $AB \neq AC$. The circle with diameter BC intersects the sides AB and AC at M and N respectively. Denote by O the midpoint of the side BC. The bisectors of the angles $BAC\angle$ and $MON\angle$ intersect at R. Prove that the circumcircles of the triangles BMR and CNR have a common point lying on the side BC.*

Solution. Since $OM = ON$, the bisector of $MON\angle$ is the perpendicular bisector of MN. Being hence the intersection of the bisector of the angle A and the perpendicular bisector of the side MN, R is the midpoint of the arc MN of the circumcircle of the triangle AMN. Denote the intersection of the line AR with the side BC by L.

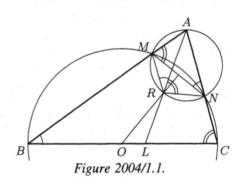

Figure 2004/1.1.

Since $BCNM$ is cyclic

$$MBC\angle = ANM\angle \quad \text{and} \quad NCB\angle = AMN\angle.$$

Similarly, in the cyclic quadrilateral $AMRN$

$$ANM\angle = ARM\angle \quad \text{and} \quad AMN\angle = ARN\angle.$$

The point in question is hence L, since

$$MBL\angle = ARM\angle \quad \text{and} \quad NCL\angle = ARN\angle$$

implies that both $BLRM$ and $CNRL$ are cyclic and thus their circumcircles are identical to those of the triangles BMR and CNR, respectively.

Since, by the condition, $AB \neq AC$, the bisectors of $A\angle$ and $MON\angle$ are not parallel, the proof is complete.

2004/2. *Find all polynomials $P(x)$ with real coefficients such that for all reals a, b, c such that $ab + bc + ca = 0$ we have the following relation*

(1) $$P(a-b) + P(b-c) + P(c-a) = 2P(a+b+c).$$

Solution. We prove that the polynomials satisfying the conditions of the problem are those of the form $P(x) = \alpha x^2 + \beta x^4$ where α and β are arbitrary real numbers. (The constant term is zero, since $P(0) = 0$.)

If $a = b = 0$ then $ab + bc + ca = 0$ holds for every real number c. The corresponding substitution yields

$$P(0-0) + P(0-c) + P(c-0) = 2P(0+0+c),$$

(2) $$P(0) + P(-c) = P(c).$$

Plugging here $c = 0$ yields $P(0) = 0$ and thus (2) reduces to

$$P(-c) = P(c).$$

The polynomial P is hence even, it can be written as $P(x) = a_n x^{2n} + a_{n-1}x^{2n-2} + \ldots + a_1 x^2$ with real numbers $a_1, \ldots, a_n$.

Now comes the tricky step: observe that if $a = 6x$, $b = 3x$ and $c = -2x$, then $ab + bc + ca = 18x^2 - 6x^2 - 12x^2 = 0$; substituting in (1):

$$P(3x) + P(5x) + P(-8x) = 2P(7x).$$

This equality holds for every real number x if and only if the coefficients of the equal powers of x are respectively equal. Therefore

(3) $$\left(3^{2n} + 5^{2n} + (-8)^{2n}\right) \cdot a_n = 2 \cdot 7^{2n} \cdot a_n.$$

If $n = 1$ then $9 + 25 + 64 = 2 \cdot 49$; if $n = 2$ then $81 + 625 + 4096 = 2 \cdot 2401$, and thus there is no restriction on the values of a_1 and a_2. For $n \geq 3$ the coefficient of a_n on the l. h. s. of (3) is definitely greater than that on the r. h. s. To see this it is enough to compare the corresponding powers of 8 and 7: even for $n = 3$

$$8^6 = 2^9 \cdot 2^9 > 500 \cdot 500 = 100 \cdot 50 \cdot 50 > 2 \cdot 49 \cdot 49 = 2 \cdot 7^6.$$

Therefore, if $n \geq 3$ then (3) holds if and only if $a_n = 0$.

So far we have seen that any solution must be of the form $P(x) = \alpha x^2 + \beta x^4$. We are left to verify if these polynomials are indeed satisfying the conditions for every value of α and β. It is clearly enough to check the particular polynomials x^2 and x^4 because the condition is linear. The proof is humble calculation: the differences of the two sides of (1) are

(4) $$(a-b)^2 + (b-c)^2 + (c-a)^2 - 2(a+b+c)^2 = -6(ab+bc+ca),$$

and

(5) $$(a-b)^4 + (b-c)^4 + (c-a)^4 - 2(a+b+c)^4 =$$

$$= -12\left[a^2(ab+bc+ca) + b^2(ab+bc+ca) + c^2(ab+bc+ca)\right] - 6(ab+bc+ca)^2,$$

respectively. By the condition $ab + bc + ca = 0$, the r. h. s. of both (4) and (5) is equal to 0, the solution is hence finished.

2004/3. *Define a* hook *to be a figure made up of six unit squares as shown below on the diagram, or any of the figures obtained by applying rotations and reflections to this figure.*

Determine all $m \times n$ rectangles that can be covered with hooks such that

- *the covering is without gaps and without overlaps,*
- *no part of a hook covers area outside the rectangle.*

Solution. Since the covering is without gaps and overlaps, there are two ways for hook B to cover the square surrounded by hook A (*Figure 2004/3.1.*). The two hooks hence fit into each other and thus the covering is realized by pairs of hooks, the area mn must be a multiple of 12.

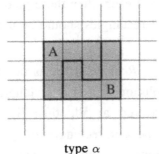

type α type β

Figure 2004/3.1.

We prove that one dimension of the rectangle has to be a multiple of 4. Assume the contrary. Since $12 \mid mn$, this can happen only if $m = 2m'$ and $n = 2n'$ with m' and n' odd numbers. The number of pairs of the hooks is equal to the third of $m' \cdot n'$, an odd number. We are going to show that this is not possible.

			1				1				1				
			1				1				1				
			1				1				1				
1	1	1	2	1	1	1	2	1	1	1	2	1	1		
			1				1				1				
			1				1				1				
			1				1				1				
1	1	1	2	1	1	1	2	1	1	1	2	1	1		
			1				1				1				
			1				1				1				

Figure 2004/3.2.

Label, for the proof, certain fields of the rectangle in the following way: write 1 into the jth square of the ith row if exactly one of i or j is a multiple of 4; write here 2 if both "coordinates" are divisible by 4 (*Figure 2004/3.2.*). Since both m and n are even and the sum of the entries is even in every 2×2 square, by construction, the total of the numbers hence entered is also even. Let's check now the sum of the numbers covered by a pair of hooks. If the pair is of type α then the covered sum is either 3 or 7 and if it is of type β then the sum is either 5 or 7, an odd number anyway. Since there are odd pairs participating in the covering altogether, the total must be also odd. This contradiction proves the claim.

Now we may assume, without loss of generality, that $3 \mid m$. If $4 \mid n$ then there is a straightforward covering by 3×4 rectangles of type α. If $12 \mid m$ then the covering is impossible if $n = 1$, 2 or 5. Any other n can be written as $n = 3k + 4l$. Accordingly, if the rectangle is now divided into k rectangles of dimensions $m \times 3$ each and l rectangles of dimensions $m \times 4$ each, then these pieces can be covered by pairs of type α, since $12 \mid m$; 4×3 rectangles should be used in the first case and 3×4 rectangles in the second.

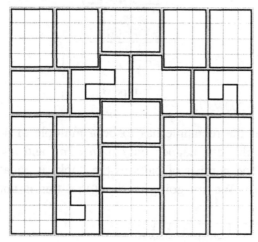

Remark. It turns out from the solution that if a rectangle can be covered at all, then this can be done with rectangles of type α only. However, this does not mean that the pairs of type β are useless: the covering on the *Figure 2004/3.3.* uses pairs of both kind.

Figure 2004/3.3.

2004/4. *Let $n \geq 3$ be an integer. Let t_1, t_2, ..., t_n be positive real numbers such that*

$$n^2 + 1 > (t_1 + t_2 + \ldots + t_n)\left(\frac{1}{t_1} + \frac{1}{t_2} + \ldots + \frac{1}{t_n}\right).$$

Show that t_i, t_j, t_k are side lengths of a triangle for all i, j, k with $1 \leq i < < j < k \leq n$.

Solution. It is enough to show, by symmetry, that $t_1 < t_2 + t_3$; the same argument clearly works for any triple t_i, t_j, t_k in arbitrary order. Expanding the

r. h. s.:

$$\sum_{i=1}^{n} t_i \sum_{i=1}^{n} \frac{1}{t_i} = n + \sum_{1 \le i < j \le n} \left(\frac{t_i}{t_j} + \frac{t_j}{t_i} \right) =$$

$$= n + t_1 \left(\frac{1}{t_2} + \frac{1}{t_3} \right) + \frac{1}{t_1}(t_2 + t_3) + \sum_{\substack{1 \le i < j \le n \\ (i,j) \ne (1,2),(1,3)}} \left(\frac{t_i}{t_j} + \frac{t_j}{t_i} \right).$$

By the AM-GM inequality

$$\frac{1}{t_2} + \frac{1}{t_3} \ge \frac{2}{\sqrt{t_2 t_3}}, \quad t_2 + t_3 \ge 2\sqrt{t_2 t_3} \quad \text{and} \quad \frac{t_i}{t_j} + \frac{t_j}{t_i} \ge 2$$

for every pair i, j.

The condition hence can be written as

$$n^2 + 1 > \sum_{i=1}^{n} t_i \sum_{i=1}^{n} \frac{1}{t_i} \ge n + 2\frac{t_1}{\sqrt{t_2 t_3}} + 2\frac{\sqrt{t_2 t_3}}{t_1} + 2 \left[\binom{n}{2} - 2 \right].$$

Rearranging and multiplying through by $t_1 \sqrt{t_2 t_3}$ (it is positive):

$$0 > 2t_1^2 + 2t_2 t_3 - 5t_1 \sqrt{t_2 t_3} = 2 \left(t_1 - \frac{1}{2}\sqrt{t_2 t_3} \right) \left(t_1 - 2\sqrt{t_2 t_3} \right).$$

The condition for this inequality to hold is clearly

$$\frac{1}{2}\sqrt{t_2 t_3} < t_1 < 2\sqrt{t_2 t_3}.$$

Estimating the second inequality by the AM–GM inequality yields

$$t_1 < 2\sqrt{t_2 t_3} \le t_2 + t_3$$

and we are done.

2004/5. *In a convex quadrilateral $ABCD$ the diagonal BD bisects neither the angle $ABC\angle$, nor the angle $CDA\angle$. A point P lies inside $ABCD$ and satisfies*

$$PBC\angle = DBA\angle \quad \text{and} \quad PDC\angle = BDA\angle.$$

Prove that $ABCD$ is a cyclic quadrilateral if and only if $AP = CP$.

Solution. First we prove that for $ABCD$ cyclic $AP = CP$.

Denote the second intersections of the lines BP and DP with the circumcircle of $ABCD$ by D' and B', respectively. Since $ABP\angle = D'BC\angle$, the intercepted arcs AD and $D'C$ are also congruent. D and D' are symmetric with respect to the perpendicular bisector of AC and the same holds for B and B'. Hence $BB'D'D$ is a symmetric trapezium and being the intersection of its diagonals P is incident to its axis. Since this line is the perpendicular bisector of AC as well, $AP = CP$, indeed.

We provide three arguments to prove the converse statement.

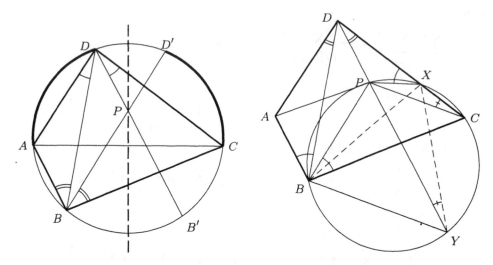

Figure 2004/5.1.   *Figure 2004/5.2.*

I. Let the circumcircle of the triangle BPC meet DC at X and DP at Y, respectively. Since $BPXC$ is cyclic, $PBC\angle = PXD\angle$. Having also two matching angles, the triangles DAB and DPX are similar (*Figure 2004/5.2.*). Therefore, $AD:DB = DP:DX$. On the other hand, $ADP\angle = BDX\angle$, and thus, by the previous equality, the triangles ADP and BDX are also similar; for the ratio of the corresponding sides

(1) $BX:AP = DX:DP.$

Being inscribed angles and intercepting the arc XP, the angles $XCP\angle$ and $XYP\angle$ are congruent. The triangles DPC and DXY are similar and the ratios of the corresponding sides are equal:

(2) $XY:CP = DX:DP.$

Since $AP = CP$, by condition, (1) and (2) imply $BX = XY$. Therefore, the angles subtended at B and P are also equal: $BCX\angle = XPY\angle$. The proof is now finished, since

$$DAB\angle + DCB\angle = DPX\angle + XPY\angle = 180°.$$

II. We shall use the theorem in [41]: since the lines DA, BA, PA passing through the vertices D, B, P of the triangle DBP, respectively, are concurrent, so are their reflections in the corresponding angle bisectors of the triangle. The reflection of DA is DC, that of BA is BC, and thus the mirror image of PA is PC. The bisector p of the exterior angle P of the triangle is also halving the angle of the lines AP and CP and thus, by $AP = CP$ the line p is the axis of the segment AC.

The line p does not contain both B and D because BD does not bisect $ABC\angle$. We may assume that it is D not incident to p; let the mirror image of B

in p be B'. Since p is the external bisector of the triangle BPD, the point B' is incident to DP. The segments AB and CB' are subtending equal angles at D, and, being symmetric in p they are congruent: $AB = CB'$. Draw now the circles passing through the endpoints of these segments corresponding to the inscribed angle ADB. These circles are symmetric in p. Since their common point, D is not incident to p, their axis of symmetry, the two circles must be identical. The point D is hence lying on the circumcircle of the symmetric trapezium $ABB'C$.

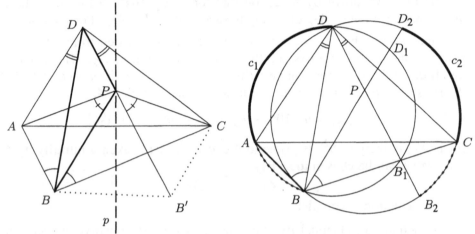

Figure 2004/5.3.   *Figure 2004/5.4.*

III. Let the circumcircles of the triangles ABD and CBD be c_1 and c_2, respectively.

The lines BP and DP meet c_1 and c_2 at D_1, D_2 and B_1, B_2, respectively (*Figure 2004/5.4.*). The corresponding arcs measured in radians of the circles c_i are denoted by $(XZ)_i$. By condition, $PBC\angle = DBA\angle$ and hence $(CD_2)_2 = (AD)_1$. Similarly, $PDC\angle = BDA\angle$ implies $(CB_2)_2 = (BA)_1$. Accordingly, there is a similarity f that changes orientation and

$$f(c_1) = c_2, \ f(A) = C, \ f(B) = B_2, \ f(D) = D_2.$$

Since $(DD_1)_1 = (DD_2)_2$ and $(BB_1)_1 = (BB_2)_2$, we have $f(D_1) = D$ and $f(B_1) = B$. Then

$$f(P) = f(BD_1 \cap B_1D) = B_2D \cap BD_2 = P.$$

Since $AP = CP = f(A)f(P)$, the scale factor of this similarity is 1, actually. Therefore, $PB_1 = f(P)f(B_1) = PB$ and $PB = f(P)f(B) = PB_2$, that is $PB_1 = PB_2$ and $B_1 = B_2$.

Since the points B, D and $B_1 = B_2$ belong to both of the circles c_1 and c_2, the two circles must be identical.

2004/6. *We call a positive integer* alternating *if every two consecutive digits in its decimal representation are of different parity.*

Find all positive integers n such that n has a multiple which is alternating.

Solution. We show that n has an alternating multiple if and only if n is not a multiple of 20. The condition is clearly necessary since if $20 \mid n$ then its last two digits are both even.

(i) First we prove by induction that for every positive integer c there is an alternating number A_{2c} of $2c$ digits such that $2^{2c+2} \mid A_{2c}$. For $c = 1$ set $A_2 = 16$ and assume that $A_{2c} = 2^{2c+2} \cdot A'_{2c}$. We are going to step from c to $c+1$ by inserting one of the numbers 16, 36, 56, 76 to the front of A_{2c}. These numbers are all divisible by 4 they can be written as $m = 4 \cdot m'$, with m' one of 4, 9, 14, 19. These numbers form a complete residue system modulo 4.

$$A_{2(c+1)} = m \cdot 10^{2c} + A_{2c} = 2^{2c+2} \left(m' \cdot 5^{2c} + A'_{2c} \right)$$

Since A'_{2c} is integer, 5^{2c} is prime to 4 and m' can be taken arbitrarily modulo 4, its value can be chosen to satisfy

$$4 \mid m' \cdot 5^{2c} + A'_{2c}$$

and the induction is complete.

(ii) An almost identical argument shows that for any positive integer d there exists an alternating number B_d of d digits such that $5^d \mid B_d$. For $d = 1$ this is $B_1 = 5$. Let, by the induction hypothesis $B_d = 5^d \cdot B'_d$. Following the previous strategy we are going to insert some number k to the front of B_d again. If d is odd then k can admit any one of the values 0, 2, 4, 6, 8, otherwise it can be equal to 1, 3, 5, 7 or 9, a complete residue system modulo 5 in both cases.

$$B_{d+1} = k \cdot 10^d + B_d = 5^d \cdot \left(k \cdot 2^d + B'_d \right).$$

Since B'_d is integer, 2^d is prime to 5 and k can be taken arbitrarily modulo 5, its value can be chosen to satisfy

$$5 \mid k \cdot 2^d + B'_d$$

It may happen that B_d starts with zero but that's all right for the construction; the next digit appended to the front is then odd and yields the next number B_{d+1}. At the final step of our construction we shall need odd digit terms of this sequence, anyway.

(iii) Let s be an arbitrary positive integer prime to 10 and M some alternating number of *even* digits. Concatenating several copies of M we obtain numbers of the form $M, \overline{MM}, \overline{MMM}, \ldots$. A standard pigeonhole argument yields that there are numbers among these M-strings whose difference

$$\underbrace{\overline{MM \ldots M}}_{q \text{ db}} \cdot 10^{2r} = \overline{M}_q \cdot 10^{2r}$$

is divisible by s. Being prime to 10 $s \mid \overline{M}_q$. Since M is an even-digit alternating number, the string formed of q copies of M is also alternating.

To complete the construction we now exhibit an alternating multiple of $n = = 2^c \cdot 5^d \cdot s$ where $(s, \, 10) = 1$ and $2^c \cdot 5^d$ is not divisible by 20. If $c = d = 0$ then any even digit alternating number will do, $M = 21$, for example. If $c \leq 1$ and $d \geq \geq 1$ then set $M = 10 \cdot B_{2d+1}$ (the number of digits in B is odd) . If $c \geq 1$ and $d = 0$ then set $M = A_{4c}$. Being an even digit alternating number in each case M is an appropriate input for the procedure *(iii)* to prepare an alternating multiple of s. The resulting multiple of M will do for our purpose since its building block, M guarantees the divisibility by $2^c \cdot 5^d$ prime to s.

7. A Glossary of Theorems

[1] *The paralelogram theorem and an application.* The sum of the squares of a paralelogram's diagonals is equal to that of the sides. Denote, for the proof, the vectors spanning the paralelogram by **a** and **b**; hence its diagonals are **a** + **b** and **a** − **b**, respectively. The claim is now straightforward as

$$2\mathbf{a}^2 + 2\mathbf{b}^2 = (\mathbf{a}+\mathbf{b})^2 + (\mathbf{a}-\mathbf{b})^2.$$

Let the lengths of the sides of a triangle be a, b and c and that of the median $CC' = s_c$ (*Figure 1.1.*). Reflecting the triangle in C' yields a paralelogram of sides a, b, a, b; its diagonals are c and $2s_c$. By the previous result

$$2a^2 + 2b^2 = 4s_c^2 + c^2,$$

and hence

$$s_c^2 = \frac{2a^2 + 2b^2 - c^2}{4}.$$

[2] *Sides and cotangents in a triangle.*

$$a^2 + b^2 + c^2 = 4t(\cot\alpha + \cot + \cot\gamma).$$

Write down the cosine rule for the sides and sum the equalities:

$$a^2 + b^2 + c^2 = 2\left(a^2 + b^2 + c^2\right) - 2bc\cos\alpha - 2ca\cos\beta - 2ab\cos\gamma,$$

$$a^2 + b^2 + c^2 = 2\left(bc\cos\alpha + ca\cos\beta + ab\cos\gamma\right).$$

From the area formula: $bc = \dfrac{2A}{\sin\alpha}$; substituting the symmetric permutations of this relation into the previous result yields the claim. Indeed

$$a^2 + b^2 + c^2 = 4t\left(\frac{\cos\alpha}{\sin\alpha} + \frac{\cos\beta}{\sin\beta} + \frac{\cos\gamma}{\sin\gamma}\right) = 4t(\cot\alpha + \cot\beta + \cot\gamma).$$

[3] *The cotangent inequality.* If α, β, γ are the angles of a triangle then

$$\cot\alpha + \cot\beta + \cot\gamma \geq \sqrt{3}.$$

There are several ways to prove this inequality; starting with straightforward identities we proceed by simple estimations.

$$\cot\alpha + \cot = \frac{\cos\alpha}{\sin\alpha} + \frac{\cos\beta}{\sin\beta} = \frac{\sin(\alpha+\beta)}{\sin\alpha\sin\beta} = \frac{2\sin\gamma}{\cos(\alpha-\beta) - \cos(\alpha+\beta)} =$$

$$= \frac{2\sin\gamma}{\cos(\alpha-\beta) + \cos\gamma} \geq \frac{2\sin\gamma}{1+\cos\gamma} = \frac{4\sin\frac{\gamma}{2}\cos\frac{\gamma}{2}}{1+2\cos^2\frac{\gamma}{2} - 1} = 2\tan\frac{\gamma}{2}.$$

Therefore

$$\cot\alpha + \cot\beta + \cot\gamma \geq \cot\gamma + 2\tan\frac{\gamma}{2} =$$

$$= 2\tan\frac{\gamma}{2} + \frac{\cot^2\frac{\gamma}{2} - 1}{2\cot\frac{\gamma}{2}} = \frac{1}{2}\frac{\cot^2\frac{\gamma}{2} + 3}{\cot\frac{\gamma}{2}} = \frac{1}{2}\left(\cot\frac{\gamma}{2} + 3\tan\frac{\gamma}{2}\right).$$

The last sum can be estimated from below by the A.M.–G.M. inequality.

$$\cot\alpha + \cot\beta + \cot\gamma \ge \sqrt{\cot\frac{\gamma}{2}\cdot 3\tan\frac{\gamma}{2}} = \sqrt{3}.$$

[4] *Isogonal point* (isogonal = equiangular). The sides of a triangle are subtending equal angles at the isogonal point. This common angle is clearly equal to $120°$ and one can find such a point if the angles of the triangle are all less than $120°$.

The isogonal point is hence incident to the circular arcs through the endpoints of the sides corresponding to $120°$; these arcs do necessarily have a point in common.

The following simple construction is also leading to the isogonal point. Draw equilateral triangles ABC', BCA' and CAB' above the sides of the triangle ABC externally; the line segments AA', BB', CC' are then congruent (*Figure 4.1.*). The segments AA' and BB', for example, are equal because the rotation by $60°$ about C is mapping the triangle ACA' into $B'CB$; this is true if these triangles are degenerated into a segment.

If the isogonal point I does exist then it is incident to each of the segments AA', BB', CC'. In fact, since $AIC\angle = AIB\angle = 120°$ rotating the triangle AIB about A by $60°$ one obtains the triangle $AI'C'$. By the same rotation the triangle AII' is equilateral (*Figure 4.2.*). The diagram reveals that the points C, I and C' are collinear, therefore I is incident to the segment CC', indeed. Apart from that $CC' = IA + IB + IC$ and thus $CC'(= AA' = BB')$ is equal to the sum of the distances of the point I from the vertices of the triangle.

Starting with a point I not lying on the segment CC' the previous rotation yields the following inequality:

$$AI + BI + CI = II' + CI + I'C' > CC',$$

since the segments II', CI, $I'C'$ now form a triangle. Differently speaking, the isogonal point – if it exists – is minimizing the sum of the distances of a point from the vertices of a triangle; this minimum is actually equal to the common length of the segments AA', BB', CC'.

When written down in the triangles BCB', CAC', ABA', respectively, the cosine rule yields

$$BB' = CC' = AA' = a^2 + b^2 - 2ab\cos(\gamma + 60°) =$$
$$= b^2 + c^2 - 2bc\cos(\alpha + 60°) = c^2 + a^2 - 2ac\cos(\beta + 60°).$$

This latter equality, by the way, can be obtained in a straightforward manner without the above investigations.

[5] *Bicentric polygons; Poncelet's porisms.* A polygon is called bicentric if it has both an inscribed and a circumscribed circle. Any triangle is bicentric, for example, and also all the regular polygons. There is a relation, for bicentric polygons,

between the inradius r, the circumradius R and the distance d of the respective centres depending also on the number of sides. Here there is a list of the first few of them for 3, 4, 5 and 6 sided polygons.

$$n = 3: \quad d^2 = R^2 - 2Rr \quad \text{or:} \quad \frac{1}{R+d} + \frac{1}{R-d} = \frac{1}{r}, \quad \text{(Euler)}$$

$$n = 4: \quad \left(R^2 - d^2\right)^2 = 2r^2 \left(R^2 + d^2\right) \quad \text{or:} \quad \frac{1}{(R+d)^2} + \frac{1}{(R-d)^2} = \frac{1}{r^2},$$

$$n = 5: \quad r(R - d) = (R + d)iR - r - d(iR - r + d + i2R),$$

$$n = 6: \quad 3\left(R^2 - d^2\right)^4 = 4r^2\left(R^2 + d^2\right)\left(R^2 - d^2\right) + 16R^2 d^2 r^4.$$

The circles of bicentric polygons have the celebrated property that they are, in fact, shared by infinitely many bicentric $n$gons; to put it more precisely denote the incircle and the circumcircle of a bicentric polygon by c and C, respectively. Draw from an arbitrary point A_1 of C a tangent to c and let this line meet C at A_2. The – other – tangent from A_2 to c meets C at A_3, etc. The point A_{n+1} of this process is then incident to A_1, the sequence of segments hence drawn is closing up at the nth step yielding the porism of Poncelet in circles.

In the general form of Poncelet's theorem there are conic sections for circles; accordingly, the general theorem is a gem of projective geometry.

[6] *Power of a point with respect to a circle; radical axis; radical centre.* Given is the circle c of radius R and centre O, for an arbitrary point P in the plane of c the real number

$$h = PO^2 - R^2$$

is called the power of the point P with respect to the circle c. This number is positive, zero or negative if the point P is outside of c, lying on c or it is inside c. For external points the value of h is equal to the square of the tangent from P to the circle.

The locus of the points whose power is equal with repect to two non concentric circles is a straight line perpendicular to their axis. This line is called the radical axis of the two circles. It is the line of the common chord if the circles intersect; it is the common – internal – tangent if they touch; in general, it contains those points from where one can draw equal tangents to the circles.

In case of three circles the pairwise drawn radical axes are either parallel or they meet at a common point; in the latter case this point is called the radical centre of the three circles.

[7] *The position vector of the incentre.* Denote the position vectors of the vertices of the triangle ABC by **a**, **b** and **c**, respectively. Since the intersection C_1 of the bisector of the angle C and the opposite side AB is dividing AB in the ratio $b:a$

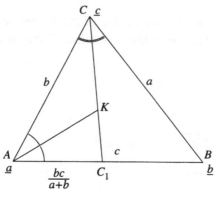

Figure 7.1.

(*Figure 7.1.*), the position vector of C_1 is equal to

$$\frac{a\mathbf{a}+b\mathbf{b}}{a+b}.$$

The bisector of the angle A meets the segment CC_1 at the incenter K dividing CC_1 in the ratio $b:\dfrac{bc}{a+b}=1:\dfrac{c}{a+b}$. K's position vector is hence

$$\mathbf{k}=\frac{\frac{c}{a+b}\cdot\mathbf{c}+\frac{a\mathbf{a}+b\mathbf{b}}{a+b}}{1+\frac{c}{a+b}}=\frac{a\mathbf{a}+b\mathbf{b}+c\mathbf{c}}{a+b+c}.$$

[8] *The inradius and the circumradius of a triangle.* There are several formulas for these radii in terms of the sides and the angles of the triangle; here there are but a few of them. The inradius is denoted by r, the circumradius is by R, the area of the triangle by A, the sides and the angles by a, b, c, and α, β, γ, respectively and finally, the semiperimeter by s.

$$R=\frac{abc}{4A}=\frac{a}{2\sin\alpha}=\frac{b}{2\sin\beta}=\frac{c}{2\sin\gamma},$$

$$R^2=\frac{2A}{\sin 2\alpha+\sin 2\beta+\sin 2\gamma}$$

$$r=\frac{A}{s}=4R\sin\frac{\alpha}{2}\sin\frac{\beta}{2}\sin\frac{\gamma}{2},$$

$$r^2=A\tan\frac{\alpha}{2}\tan\frac{\beta}{2}\tan\frac{\gamma}{2}.$$

[9] *Poncelet's theorem* see [5].

[10] *Touching circles.* Two circles in the space are touching each other if they have a single point in common and they share the tangent at this point. The axis of a circle is the line perpendicular to its plane at the centre. The axis contains the points whose distance from the points of the circle are all equal.

The plane perpendicular to the common tangent at the point T_{12} of contact of the touching circles c_1 and c_2 is containing the axes of both; if the circles are not coplanar then the axes meet and the distance of their common point O is equal to OT_{12} from the points of both circles; they are hence incident to the sphere of centre O and radius OT_{12}.

Consider now three pairwise touching circles c_1, c_2, c_3 that are not coplanar. Denote the touching points by T_{12}, T_{13}, T_{23}, respectively. Let the sphere containing c_1 and c_2 be S. Since there is a unique sphere passing through a circle c and a point not lying on c, the circle c_1 and the point T_{23} determine the sphere S; this sphere is also passing through the circle c_3. Accordingly, three pairwise touching circles, if not coplanar, are lying on a sphere.

[11] *Equilateral tetrahedron* A tetrahedron is called equilateral if its faces are congruent. Being the face of an equilateral tetrahedron a triangle is always acute angled; any such triangle, on the other hand, is the face of some equilateral tetrahedron.

The opposite edges of an equilateral tetrahedron are pairwise equal; its circumscribed parallelepiped is hence a cuboid. Its specific points, namely the incentre, the circumcentre and its centroid are incident; conversely, if any two of the above points are incident then the tetrahedron is equilateral. Finally, if the areas of the faces of a tetrahedron are equal then the terahedron is equilateral.

[12] *The cosine inequality.* As a fundamental inequality in triangles it is the source of several further results; it states that

$$\cos\alpha + \cos\beta + \cos\gamma \le \frac{3}{2},$$

and equality holds if and only if the triangle is equilateral.

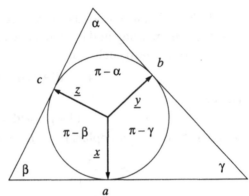

Figure 12.1.

The following non standard argument is using vectors: set the origin as the incentre and let the inradius be equal to 1. The unit vectors to the points of contact of the incircle are denoted by **x**, **y**, **z**, respectively. Using the notations of *Figure 12.1*

$$0 \le (\mathbf{x}+\mathbf{y}+\mathbf{z})^2 = 3 + 2\mathbf{xy} + 2\mathbf{yz} + 2\mathbf{zx} =$$

$$= 3 + 2\left(\cos(a\pi - \gamma) + \cos(a\pi - \alpha) + \cos(a\pi - \beta)\right) = 3 - 2\left(\cos\alpha + \cos\beta + \cos\gamma\right),$$

yielding

$$\cos\alpha + \cos\beta + \cos\gamma \le \frac{3}{2},$$

and equality holds if and only if $\mathbf{x}+\mathbf{y}+\mathbf{z}=\mathbf{0}$, that is $1 = \mathbf{x}^2 = (-\mathbf{y}-\mathbf{z})^2 = 2 + 2\cos(a\pi - \alpha)$, $\cos(a\pi - \alpha) = -\frac{1}{2}$, $180° - \alpha = 120°$. The pairwise angles of the vectors **x**, **y**, **z** are $120°$, the triangle is equilateral.

[13] *Ramsey's theorem.* If the edges of an n-point graph are coloured with two colours say blue and red then Ramsey's theorem claims the existence of a monochromatic complete subgraph of certain size. Formally:

For every pair (b, r) of positive integers there exists a positive integer $R(b, r)$ such that if $n \ge R(b, r)$ and each edge of a complete n-graph is coloured either blue or red then either there is a complete subgraph of b vertices whose edges are all blue or there is a complete subgraph of r vertices whose edges are all red. If, on the other hand, $n < R(b, r)$, then one can colour the edges of a complete

n-graph in such a way that there are no monochromatic complete subgraphs of the given sizes.

The task of finding the actual values of the so called Ramsey numbers $R(b, r)$ is extremely hard in general. There are some estimations but they are far from being sharp.

[14] *Position vectors of coplanar points.* Let the vectors **a, b, c, d** start from the origin, otherwise be arbitrary. If **a, b** and **c** are not coplanar then the endpoints of the quadruple **a, b, c, d** are coplanar if and only if there exist real numbers α, β, γ such that

(1) $$\mathbf{d} = \alpha\mathbf{a} + \beta\mathbf{b} + \gamma\mathbf{c}, \qquad \alpha + \beta + \gamma = 1.$$

Indeed, if the endpoints of the vectors are coplanar then the vectors $\mathbf{a} - \mathbf{d}$, $\mathbf{b} - \mathbf{d}$, $\mathbf{c} - \mathbf{d}$ are lying in the same plane and, accordingly, there exist real numbers λ, γ, ν not all zero such that

$$\lambda(\mathbf{a} - \mathbf{d}) + \gamma(\mathbf{b} - \mathbf{d}) + \nu(\mathbf{c} - \mathbf{d}) = \mathbf{0}.$$

Hence

$$(\lambda + \mu + \nu)\mathbf{d} = \lambda\mathbf{a} + \mu\mathbf{b} + \nu\mathbf{c}.$$

Since **a, b, c** are not coplanar, $\lambda + \mu + \nu \neq 0$, and thus with

$$\alpha = \frac{\lambda}{\lambda + \mu + \nu}, \qquad \beta = \frac{\mu}{\lambda + \mu + \nu}, \qquad \gamma = \frac{\nu}{\lambda_\mu + \nu}$$

we get the desired result. Since the steps of the argument can be reversed the proof is complete.

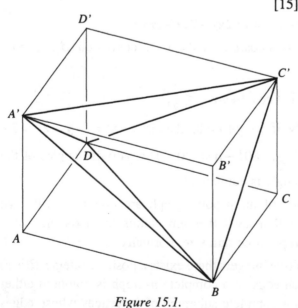

Figure 15.1.

[15] *Circumscribed parallelepiped.* The endpoints of the non parallel diagonals of two opposite (parallel) faces of a parallelepiped fom a tetrahedron. The edges of this tetrahedron are the face diagonals of the prism and opposite edges are lying on oppposite faces. The prism itself is the circumscribed parallelepiped of the tetrahedron. (*Figure 15.1*)

One can, in fact, construct a circumscribed parallelepiped about any tetrahedron by laying a plane through each edge parallel to the opposite edge; the circumscribed parallelepiped of the tetrahedron is bounded by the planes hence obtained.

Certain properties of the tetrahedron become transparent when one takes the circumscribed parallelepiped into account. The centroid of a tetrahedron, for example, is incident to the centre of this parallelepiped. The circumscribed parallelepiped of an equilateral tetrahedron is a cuboid, since as the opposite edges of the tetrahedron, the diagonals are equal on each face.

[16] *Hero's formula.* Hero's formula expresses the area of a triangle in terms of its sides. There are several ways to draft it:

$$A^2 = s(s-a)(s-b)(s-c) = \frac{1}{16}(a+b+c)(-a+b+c)(a-b+c)(a+b-c) =$$

$$= \frac{1}{16}\left[(a+b)^2 - c^2\right]\left[c^2 - (a-b)^2\right] = \frac{1}{16}(2a^2b^2 + 2b^2c^2 + 2c^2a^2 - a^4 - b^4 - c^4) =$$

$$= \frac{1}{16}\begin{vmatrix} 0 & 1 & 1 & 1 \\ 1 & 0 & c^2 & b^2 \\ 1 & c^2 & 0 & a^2 \\ 1 & b^2 & a^2 & 0 \end{vmatrix}.$$

[17] *Specific points of an equilateral tetrhedron*   see [11].

[18] *Convex hull.* The convex hull of a set S of points is the convex set containing S and contained by each convex set that contains S. To put it simply the convex hull is the "smallest" convex set containing S,

There is a nice way to visualize the convex hull of a finite planar set: imagine that the points are nails driven in a table and stretch an elastic ribbon about them: the convex hull is the region bounded by the elastic.

Convex hulls in the plane (in the space) can be obtained as the intersections of halfplanes (halfspaces) containing the set S.

[19] *Tangent segments in a triangle.* A frequently used fact in elementary arguments that the points of contacts of the circles that are touching the sides of a triangle are dividing them into segments whose lengths can be expressed in a simple manner in terms of the sides of the triangle. The underlying elementary fact is that the tangents to a circle from an external point are congruent. The review of their respective lengths can be checked on the *Figures 19.1* and *19.2*.

[20] *Orthocentric tetrahedron.* The lines from the vertices of a tetrahedron perpendicular to the opposite faces are the altitudes of the tetrahedron. If they meet at a common point then this point is called the orthocentre of the tetrahedron and the tetrahedron itself is orthocentric.

Here there are but a few properties of orthocentric tetrahedra:

a) the opposite edges are perpendicular;

b) the sum of the squares is the same for each pair of opposite edges;

c) the faces of their circumscribed parallelepiped are rhombs;

d) their centroid, orthocentre and the centre of the circumscribed sphere are collinear (Euler line).

Conditions a), b), c) are also sufficient for a tetrahedron to be orthocentric; moreover, it is enough to assume that the condition in either a) or b) holds for two opposite pairs of edges only.

[21] *Euler's totient function; Euler's congruence theorem.* The number of the non negative integers up to m that are coprime to m is is the totient function of Euler; its value is denoted by $\varphi(m)$. Here there are a few of its important properties:

1. $a^{\varphi(m)} \equiv 1 \pmod{m}$ if a is a positive integer and $(a, m) = 1$. This is Euler's congruence theorem.

2. If $(a, b) = 1$ then $\varphi(ab) = \varphi(a) \cdot \varphi(b)$.

3. If m is written as the product primes: $m = p_1^{\alpha_1} p_2^{\alpha_2} \dots p_r^{\alpha_r}$ then,

$$\varphi(m) = m \left(1 - \frac{1}{p_1}\right) \left(1 - \frac{1}{p_2}\right) \dots \left(1 - \frac{1}{p_r}\right) \quad \text{and } \varphi(1) = 1.$$

[22] *Cauchy's inequality.* Let $(a_1, a_2, \dots, a_n)$ and $(b_1, b_2 \dots, b_n)$ be n-tuples of real numbers ("n-dimensional vectors"); then

$$(a_1 b_1 + a_2 b_2 + \dots + a_n b_n)^2 \le \left(a_1^2 + a_2^2 + \dots + a_n^2\right) \left(b_1^2 + b_2^2 + \dots + b_n^2\right).$$

This inequality is often stated as

$$a_1 b_1 + a_2 b_2 + \dots + a_n b_n \le \sqrt{a_1^2 + a_2^2 + \dots + a_n^2} \sqrt{b_1^2 + b_2^2 + \dots + b_n^2}.$$

If the numbers b_i are not all zero then equality holds if and only if there exists a real number $\lambda \ne 0$ such that $a_i = \lambda b_i$ $(i = 1, 2, \dots, n)$ (The n-dimensional vectors are "parallel").

Its simplest proof is as follows: if the numbers b_i are not all zero then the quadratic

$$(a_1 - \lambda b_1)^2 + (a_2 - \lambda b_2)^2 + \dots + (a_n + \lambda b_n)^2 = 0$$

in λ has a solution if $a_i = \lambda b_i$ (for every i). Since there can be no more than one solution, its discriminant is not positive and this is exactly the claim. If, on the other hand, the numbers b_i are all zero, then the inequality is obvious.

[23] *Groups.* One of the most frequently occuring algebraic structures. A set of elements forms a group if there is a law of composition which when acting on arbitrary two elements on a definite order assigns an element of the set to this pair. This operation is usually referred to as group multiplication and it is denoted as the ordinary multiplication of numbers. If, for example, a and b are two elements of the group then ab is their product. This operation has the following properties:

1. If a, b, c are the elements of the group then $(ab)c = a(bc)$ (associativiy);

2. There exists an element e of the group such that for any element a

$$ea = ae = a.$$

(e is the neutral element of the group);

3. For any element a of the group there exists an element denoted by a^{-1} such that

$$aa^{-1} = a^{-1}a = e.$$

a^{-1} is the inverse of a.

If $ab = ba$ holds for any two elements, the operation is commutative, then the group is called abelian and the binary operation is then called – and denoted – addition.

Examples

1. The set of integers under addition. Zero is the neutral element and the inverse of each integer is its opposite.

2. The set of real numbers when zero is excluded under multiplication. The neutral element is 1 and the inverse of every element is its reciprocal.

3. The rotations mapping a regular hexagon into itself. The operation is the composition of rotations. The neutral element is the identity transformation and the inverse of the rotation by α is the rotation by $-\alpha$. This is a finite non abelian group.

[24] *Ptolemy's theorem.* In its general form it states that for the opposite sides a, c and b, d and the diagonals e, f of a convex quadrilateral

$$ac + bd \geq ef,$$

and equality holds if and only if the quadrilateral is cyclic. In the proof we shall refer to the notations of *Figure 24.1*.

Apply, for the triangle DAB a rotation and enlargement about D which maps it into the triangle DCB'. The scale factor of the enlargement is $\dfrac{c}{d}$ and thus $DB' = \dfrac{ec}{d}$ and $CB' = \dfrac{ac}{d}$. The triangles ADC and BDB' are similar because they are mathcing in their angle at D and also in the ratio of the neighbouring sides; the scale factor of this similarity is $\dfrac{e}{d}$ and hence $BB' = \dfrac{f \cdot e}{d}$.

The triangle inequality for the triangle BCB' yields

$$b + \frac{ac}{d} \geq \frac{ef}{d}, \qquad \text{that is} \qquad ac + bd \geq ef.$$

Equality holds if and only if C is lying on the segment BB' that is $\alpha + \gamma = 180°$, the quadrilateral is indeed cyclic.

[25] *Equilateral cones.* A circular cone is called equilateral if it has three pairwise perpendicular generators.

We are going to prove the following theorem:

If a circular cone has three pairwise perpendicular generators then there are infinitely many such triples, moreover, every generator is the member of such a perpendicular triple.

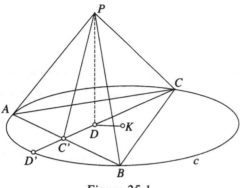

Figure 25.1.

Denote the apex of the cone by P and its base circle by c.

Let the three pairwise perpendicular generators be PA, PB, PC and denote the perpendicular projection of P on the plane of c by D (*Figure 25.1.*). It is easy to prove (Pithagoras, for example) that the triangle ABC is acute angled. Next we show that D is the orthocentre of the triangle ABC. This would follow if the line connecting D to any vertex of the triangle is perpendicular to the opposite side. Consider the vertex C, for example. The side AB is perpendicular to PC because the latter is perpendicular to the plane ABP (it is, in fact, perpendicular to two straight lines lying in this plane) and thus AB is perpendicular to every straight line in the plane ABP. On the other hand, AB is perpendicular to PD since the latter is perpendicular to the plane of c. Therefore, AB is perpendicular to two intersecting lines both lying in the plane PDC and thus it makes a right angle with any line in this plane, CD in particular, therefore, CD is an altitude, indeed.

Denote the intersection of CD with AB by C' and its second intersection with the circle c by D'. Since the mirror image of the orthocentre in a side is incident to the circumcircle, $DC' = C'D'$. In the right triangle $C'PC$ the altitude to the hyptenuse is $PD = h$ and thus, by the geometric mean theorem

$$(1) \qquad PD^2 = h^2 = C'D \cdot DC = \frac{1}{2}D'D \cdot DC.$$

The product $DD' \cdot DC$ is the power of the point D with respect to the circle c. (See [6].) This product is hence independent of C's choice.

Choose now an arbitrary point C_1 on the circle and let C_1D meet the circle at D_1. The perpendicular bisector of DD_1 meets the circle at A_1 and B_1 and the midpoint of DD_1 is C_1'. We show that the generators PA_1, PB_1 and PC_1 are pairwise perpendicular. Since

$$C'D \cdot DC = C_1D \cdot DC_1,$$
$$PD^2 = C_1'D \cdot DC_1,$$

by (1), the triangle $C_1'PC_1$ is right angled (by the converse of the geometric mean theorem). Since A_1B_1 is perpendicular to both C_1D and PD, it is also perpendicular to their plane and hence to PC_1 and thus, by the same way, since PC_1 is perpendicular to PC_1', it is also perpendicular to PA_1 and PB_1.

We are left to prove that PA_1 and PB_1 are also perpendicular. Note first, for the proof, that D has to be the orthocentre of the triangle $A_1B_1C_1$. Indeed, the orthocentre is the only one point on the altitude from C_1 whose mirror image in the side is lying on the circumcircle and by the construction of the points A_1B_1 the point D now has this property. Similarly, we obtain that PA_1 is perpendicular to PC_1 and also to PB_1. We have thus shown that any generator of the cone belongs to some family of pairwise perpendicular triple of generators.

Circular cones of this property are called equilateral. It follows from the proof that for a given circle c and point P whose distance from the plane of the circle is h and whose perpendicular projection on the base plane is D the point P is the apex of an equilateral cone of directrix c if and only if PD^2 is equal to the half of the power of D with respect to c.

In order to prove that in problem No. 2 of 1978 there are in fact infinitely many cuboids whose vertex opposite to P is Q we have to show that for a given point P and a circle c the vectors $\overrightarrow{PA}$, $\overrightarrow{PB}$, $\overrightarrow{PC}$ are spanning a cuboid whose vertex opposite to P is Q. This would follow had we shown that the sum $\overrightarrow{PA} + \overrightarrow{PB} + \overrightarrow{PC}$ is constant.

Let the vector from the centre K of c to D be $\mathbf{d}$ (this vector is clearly constant). Now

$$\overrightarrow{PA} + \overrightarrow{PB} + \overrightarrow{PC} = 3\overrightarrow{PD} + \overrightarrow{DA} + \overrightarrow{DB} + \overrightarrow{DC} =$$
$$= 3\overrightarrow{PD} + \left(\overrightarrow{KA} - \mathbf{d}\right) + \left(\overrightarrow{KB} - \mathbf{d}\right) + \left(\overrightarrow{KC} - \mathbf{d}\right) =$$
$$= 3\overrightarrow{PD} - 3\mathbf{d} + \left(\overrightarrow{KA} + \overrightarrow{KB} + \overrightarrow{KC}\right).$$

It is well known that the sum of the vectors from the circumcentre to the vertices of a triangle is leading to the orthocentre. Therefore, $\overrightarrow{KA} + \overrightarrow{KB} + \overrightarrow{KC} = \mathbf{d}$ and thus

$$\overrightarrow{PA} + \overrightarrow{PB} + \overrightarrow{PC} = 3\overrightarrow{PD} - 2\mathbf{d},$$

and this sum does not depend on the choice of the points A, B, C indeed.

[26] *A representation of positive integers.* In the solution of the problem No. 3. of 1978 we have used the following theorem: if α and β are positive irrational numbers satisfying $\dfrac{1}{\alpha} + \dfrac{1}{\beta} = 1$ then the sequences

$$\{[n\alpha]\}, \qquad [\{n\beta\}] \qquad n = 1, 2, \ldots$$

have no common elements and together they exhibit every positive integer.

Note first that the two numbers α and β are greater than 1 and thus the sequences $[n\alpha]$, $[n\beta]$ are strictly increasing. We now prove that, for any positive integer N there is either a positive integer k such that $[k\alpha] = N$, or a positive integer m such that $[n\beta] = N$, moreover, the two options cannot hold at the same time. There clearly exist the unique positive integers k and m such that

$$[(k-1)\alpha] < N \leq [k\alpha], \qquad \text{and}$$
$$[(m-1)\beta)] < N \leq [m\beta].$$

Therefore,

$$k\alpha - \alpha < N < k\alpha,$$
$$m\beta - \beta < N < m\beta.$$

Subtracting N from each term of the above inequalities:

$$(k\alpha - N) - \alpha < 0 < k\alpha - N,$$
$$(m\beta - N) < 0 < m\beta - N.$$

Introducing the notations $d = k\alpha - N$ and $d' = m\beta - N$ we get

(1) $\qquad 0 < d < \alpha, \quad 0 < d' < \beta, \quad \text{that is} \quad 0 < \dfrac{d}{\alpha} < 1, \quad 0 < \dfrac{d'}{\beta} < 1.$

Since $k = \dfrac{N}{\alpha} + \dfrac{d}{\alpha}$, $m = \dfrac{N}{\beta} + \dfrac{d'}{\beta}$, (1) implies

$$k + m = N\left(\frac{1}{\alpha} + \frac{1}{\beta}\right) + \frac{d}{\alpha} + \frac{d'}{\beta} = N + \frac{d}{\alpha} + \frac{d'}{\beta},$$

that is

(2) $\qquad\qquad\qquad\qquad \dfrac{d}{\alpha} + \dfrac{d'}{\beta} = k + m - N.$

By (1), on the other hand

$$0 < k + m - N < 2.$$

Since k, m, N are positive integers, this implies

$$k + m - N = 1.$$

Now by (2)

$$\frac{d}{\alpha} + \frac{d'}{\beta} = 1 = \frac{1}{\alpha} + \frac{1}{\beta},$$

that is $\qquad\qquad\qquad\qquad \alpha(d' - 1) = (1 - d)\beta.$

Since α and β are positive and d is irrational, this equality implies that one of d and d' is less than 1 and the other one is greater than 1. If, for example, $d < 1$, $d' > 1$ then, by the definition of d and d' implies

$$\alpha k = N + d, \qquad\qquad [\alpha k] = N,$$
$$\beta m = N + d' \qquad\qquad [\beta m] > N,$$

and thus N belongs to exactly one of the sequences $[\alpha k]$ and $[\beta m]$. The same holds if $d > 1$ and $d' < 1$.

[27] *Solving linear recurrences.* To find a formula for the nth term of a sequence defined by recurrence relations is always an important task. Here we present the solution of the following special case of the problem.

$$(1) \qquad\qquad a_n = c_1 a_{n-1} + c_2 a_{n-2}$$

This is a so called second order linear recurrence, the coefficients a_1 and a_2 are given numbers.

　　The heart of the matter is to find geometric progressions satisfying (1) and unfold the general solution as a linear combination of these particular sequences. Here there is the method. The quadratic

$$x^2 - c_1 x - c_2 = 0$$

is called the *characteristic equation* to the recurrence (1). Denote its roots (real or complex) by x_1 and x_2.

　　Assume, first, that $x_1 \neq x_2$. Then the nth term of the sequence is equal to

$$(2) \qquad\qquad a_n = \lambda x_1^n + \mu x_2^n,$$

where λ and μ are constants depending on the initial terms of the sequence; their actual value can be computed by solving the simultaneous system

$$(3) \qquad\qquad \begin{aligned} \lambda x_1 + \mu x_2 &= a_1, \\ \lambda x_1^2 + \mu x_2^2 &= a_2. \end{aligned}$$

　　If $x_1 = x_2$ then the nth term can be computed as

$$a_n = \lambda x_0^2 + \mu n x_0^{n-1}$$

(we have adopted the notation $x_1 = x_2 = x_0$). The system for the values of λ and μ is now

$$\lambda x_0 + \mu = a_1,$$

$$\lambda x_0^2 + 2\mu x_0 = a_2.$$

　　The method works essentially the same way in the general case. Consider the sequence $\{a_i\}$ defined by

$$(4) \qquad\qquad a_{n+k} = c_1 a_{n+k+1} + c_2 a_{n+k+2} + \ldots + c_k a_n.$$

This relation is called k-order linear recurrence with constant coefficients. The numbers c_i are constants and there are also given the initial values $a_1, a_2, \ldots, a_k$. The characteristic polynomial corresponding to the above recurrence is

$$x^k - c_1 x^{k-1} - c_2 x^{k-2} - \ldots - c_k = 0.$$

　　If its roots (real or complex) are $x_1, x_2, \ldots, x_k$ are distinct then the terms of the sequence $\{a_i\}$ can be computed as

$$(5) \qquad\qquad a_n = \lambda_1 x_1^{n-1} + \lambda_2 x_2^{n-1} + \ldots + \lambda_n x_k^{n-1}$$

where the coefficients $\lambda_1, \lambda_2, \ldots, \lambda_k$ can be calculated from the initial values of the sequence.

The method still works if there happen to be multiple roots of the characteristic polynomial, although, as in the second order case, the actual solution is a bit more tedious.

[28] *Two relations among binomial coefficients.*

A) $\dbinom{n}{k} = \dbinom{n-1}{k} + \dbinom{n-1}{k-1};$

B) $\dbinom{n+1}{k+1} = \dbinom{n}{k} + \dbinom{n-1}{k} + \ldots + \dbinom{k}{k}.$

The proof of A) is straightforward from the defining equality $\dbinom{n}{k} =$

$$= \frac{n!}{k!(n-k)!}.$$

$$\binom{n-1}{k} + \binom{n-1}{k-1} = \frac{(n-1)!}{k!(n-k-1)!} + \frac{(n-1)!}{(k-1)!(n-k)!} =$$

$$= \frac{(n-1)!(n-k+k)}{k!(n-k)!} = \binom{n}{k}.$$

To prove B) one can use A):

$$\binom{n+1}{k+1} = \binom{n}{k+1} + \binom{n}{k},$$

$$\binom{n}{k+1} = \binom{n-1}{k+1} + \binom{n-1}{k},$$

$$\binom{n-1}{k+1} = \binom{n-2}{k+1} + \binom{n-2}{k},$$

$$\vdots$$

$$\binom{n-(n-k-2)}{k+1} = \binom{k+2}{k+1} = \binom{k+1}{k+1} + \binom{k+1}{k}.$$

Summing the equalities and considering that $\dbinom{k+1}{k+1} = \dbinom{k}{k}$ one arrives to the claim.

[29] *Menelaus' theorem.* Let C_1, A_1 and B_1 be points on the sides AB, BC, CA of the triangle ABC, respectively. These points are collinear if and only if

(1) $$\frac{AC_1}{C_1B} \cdot \frac{BA_1}{A_1C} \cdot \frac{CB_1}{B_1A} = -1.$$

As for the sign of the fractions on the l. h. s. they are positive if the vectors $\overrightarrow{AC_1}$ and $\overrightarrow{C_1B}$ are oriented similarly, otherwise they are negative.

Note here that disregarding the orientation of the segments there is 1 on the r. h. s. of (1) and this form of the claim is just necessary for the points A_1, B_1, C_1 to be collinear.

[30] *Residue classes, congruences.* For a given integer $m > 1$ the integers a and b are said to belong to the same residue class "with respect to m", or simply *modulo m* ("mod m", for short) if they give equal remainders when divided by m. Since the possible remainders are 0, 1, 2, ..., $m - 1$, there are m residue classes with respect to m and every integer belongs to exactly one of them, or, putting it differently, every whole number is representing some residue class.

m integers form a so called *complete residue system* mod m if they represent distinct residue classes; together they hence represent every possible remainder mod m.

Two integers clearly belong to the same residue class mod m if their difference is divisible by m. For given integers a and b this is denoted as

$$a \equiv b \pmod{p} \qquad \text{or simply} \qquad a \equiv b \quad (m).$$

This is the relation of congruence. Several properties of this relation are resembling to those of equality; here there are a few of them. (For sake of brevity the mod m extension is now omitted.)

1. if $a \equiv b$, akkor $b \equiv a$; $a \equiv a$ for every integer a;

2. if $a \equiv b$ and $b \equiv c$ then $a \equiv c$;

3. if $a \equiv b$ and $c \equiv d$ then

$$a + c \equiv b + d, \quad a - c \equiv b - d, \quad ac \equiv bd, \quad a^n \equiv b^n, \quad (n \text{ is a positive integer});$$

4. if $ac \equiv bc$ then $a \equiv b \left(\text{mod } \dfrac{m}{d} \right)$, where $d = (c, m))$.

[31] *Designs.* The v element set H is called block design if there exist b subsets in H of k elements each – the blocks – every element of H belongs to exactly r blocks and any two distinct elements of H belong to exactly λ blocks.

The numbers v, b, k, r, λ are the *parameters* of the block design and they are related in several ways. (Obviously $2 \le k < v$ and $\lambda > 0$). Assign, to the elements of the block design the rows of a matrix of v rows and b columns and its columns to the blocks as follows: a given entry of the matrix is equal to 1 if the element corresponding to its row belongs to the block corresponding to its column; otherwise the entry is equal to zero. The matrix hence obtained is the so called incidence matrix of the block design. There are exactly r copies of 1 in each row and k 1-s in each column. Tallying the 1-s both row and columnwise yields

$$bk = vr.$$

Counting further incidences one gets

$$r(k-1) = \lambda(v-1).$$

To find the proper conditions under which there exists a block design for a given system of parameters is one of the unsolved hard problems of combinatorics. If, for example, $v = b = p^{2\alpha} + p^{\alpha} + 1$, $\lambda = 1$, $k = r = p^{\alpha} + 1$, where p is a prime and α is a positive integer then one can construct the corresponding block design, these are the so called finite projective planes.

[32] *Fermat's congruence theorem (the little "Fermat's theorem").* For any prime p the number of coprimes to p below p is equal to $p-1$, $\varphi(p) = p-1$ and by Euler's congruence theorem [21]

$$a^{p-1} \equiv 1 \pmod{p}, \qquad \text{if } (a, p) = 1.$$

This is Fermat's congruence theorem. Multiplying both sides by a yields

$$a^p \equiv a \pmod{p}.$$

This form of the theorem holds even if a is not prime to p since as a prime, p then divides a.

[33] *Erdős–Mordell inequality.* Denote the distances of a point P of a triangle from the vertices by p, q and r, respectively, and the perpendicular distances of P from the sides by x, y and z, respectively. Then one has the following inequality

$$p + q + r \geq 2(x + y + z)$$

and equality holds if and only if the triangle is equilateral and P is its centre. This is the Erdős–Mordell-inequality and we are going to prove it.

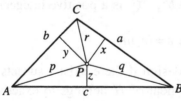

Figure 33.1.

Assume that p, q, r are the distances from the vertices A, B, C and, x, y, z are those from the sides BC, CA, AB, respectively (*Figure 33.1.*).

First we prove the following inequalities:

(1) $ap \geq bz + cy$; $\quad bq \geq cx + az$; $\quad cr \geq zy + bx$.

It is clearly enough to show the first one. Reflect, for the proof, the vertices B and C in the bisector of the angle A and denote the mirror images by C' and B', respectively; the point P remains fixed. The sides of the triangle $AB'C'$ are now $AB' = c$, $B'C' = a$, $C'A = b$; the distances of P from the sides $B'C'$, $C'A$, AB' are x', y, z, respectively (*Figure 33.2.*). The quantity x' refers to signed distance: it is zero, if P is incident to $B'C'$ and if P is outside the triangle $A'B'C'$ then x' is negative.

Denote the altitude from A of the triangle $AB'C'$ by h_a; this is clearly not longer than the path $p+x'$ from A to $B'C'$ through P. Hence

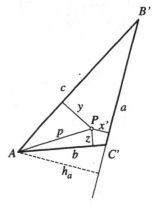

(2) $$m_a \leq p+x'.$$

Multiplying both sides by a and considering that both ah_a and $ax'+bz+cy$ are equal to the double area of the triangle $AB'C'$

$$am_a \leq ap+ax';$$

$$ax'+bz+cy \leq ap+ax',$$

Figure 33.2.

and this implies the claim $ap \geq bz+cy$; a similar argument yields the other two inequalities in (1). Note that the relations hold even if $x' \leq 0$.

Rearranging the inequalities just proved:

$$p \geq \frac{b}{a}z + \frac{c}{a}y;$$

$$q \geq \frac{c}{b}x + \frac{a}{b}z;$$

$$r \geq \frac{a}{c}y + \frac{b}{c}x.$$

The sum of these inequalities gives

(3) $$p+q+r \geq \left(\frac{c}{b}+\frac{b}{c}\right)x + \left(\frac{c}{a}+\frac{a}{c}\right)y + \left(\frac{b}{a}+\frac{a}{b}\right)z.$$

Since the sum of a positive number and its reciprocal is at least 2, one arrives to

$$p+q+r \geq 2(x+y+z).$$

As for equality it must also hold in both (2) and (3). In the latter it holds if and only if $a=b=c$; the sum of a positive number and its reciprocal is 2 only if this number is equal to 1. Our triangle hence must be equilateral.

If this is the case, however, then the triangles ABC and $AB'C'$ in the first paragraph are identical. Hence the position of the segments AP and x' must be identical to that of AP and x; finally, h_a is equal to $(p+x)$ if P is incident to the altitude from A.

This, of course, must hold for the other two altitudes, as well and thus, in case of equality, P must be the intersection of the altitudes, the centre of the equilateral triangle.

[34] *Brocard points.* The points Q_1 and Q_2 are called the Brocard points of the triangle ABC if

$$BAQ_1\angle = CBA_1\angle = ACQ_1\angle, \quad \text{or} \quad ABQ_2\angle = BCQ_2\angle = CAQ_2\angle.$$

These points can be constructed; Q_1, for example, is the intersection of two circles: one of them is passing through A and B and it is touching the line BC and the other one is passing through B and C and touching the line AC. The angles at Q_1 and Q_2 are also equal to each other and

$$\cot\omega = \cot\alpha + \cot\beta + \cot\gamma$$

for their common measure ω. (This is straightforward from the sine rule, for example.)

Apart from he circumcentre of a triangle H the Brocard points are those that have the following property: the feet of the perpendiculars to the sides from these points form a triangle similar to H. This implies that no matter how we rotate the triangle about any one of its Brocard points, the pairwise intersections of the corresponding sides of the two triangles – H and the rotated one – form a triangle that is also similar to H.

[35] *A common origin of certain inequalities.* There is a fundamental inequality showing up in the solutions from time to time. It has various formulations and there are several further inequalities that can be deduced from it. It states that

let $a_1, a_2, \ldots, a_n$ and $b_1, b_2, \ldots, b_n$ are real numbers and $b_{i_1}, b_{i_2}, \ldots, b_{i_n}$ is an arbitrary rearrangement of the numbers b_i. Prepare the sum

$$S = a_1 b_{i_1} + a_2 b_{i_2} + \ldots + a_k b_{i_k}.$$

This sum is maximal if and only if the ordering of the numbers a_i and b_{i_k} is the same and it is minimal if the two orderings are opposite.

Denote, for he proof, the highest terms of the two n-tuples by a_r and b_s, respectively and consider the sum

$$Q = a_1 b_1 + \ldots + a_r b_r + \ldots + a_s b_s + \ldots + a_n b_n.$$

Swap now the two numbers b_r and b_s; if $r \neq s$ then

$$Q' = a_1 b_1 + \ldots + a_r b_s + \ldots + a_s b_r + \ldots + a_n b_n.$$

$$Q - Q' = a_r b_s + a_s b_r - a_r b_r - a_s b_s = (a_r - a_s)(b_s - b_r) \geq 0.$$

$Q' = Q$ holds if and only if $a_r = a_s$ or $b_r = b_s$, but then the highest a_i is, in fact, multiplied by the highest b_i. Through an appropriate sequence of swaps one arrives to similarly ordered n-tuples. Since the sum Q is not decreasing, the maximum is attained when the orderings are the same, indeed. The corresponding statement about the minimum follows similarly.

[36] *Four circles theorem.* Consider four straight lines whose pairwise intersections are distinct. The circumcircles of the four triangles hence obtained are passing through a common point S (*Figure 36.1.*)

Denote, as in the diagram, the intersection of the circumcircles of the triangles ABC and CDE by S. It is clearly enough to show, by symmerty, that the circumcircle of the triangle ADF is, in fact, passing through S. Intercepted by the same arcs $SDE\angle = SCE\angle$ and in the cyclic quadrilateral $ABCS$ this angle is equal to $SAF\angle$, the quadrilateral $ASDF$ is cyclic and thus the circumcircle of the triangle ADF is passing through S, indeed.

By the Simson theorem the feet of the perpendiculars from S to the four lines are collinear. Of what we know about conic sections, it follows that four lines as tangents to the curve determine a unique parabola; the feet of the perpendiculars from the focus to the tangents are on the tangent through the vertex; the circumcircles of the triangles formed by three tangents to the parabola are passing through the focus. Taking these facts into account we get that the point S is, in fact, the focus of the parabola determined by the four straight lines.

[37] *Radius inequality.* The diameter of the incircle of a triangle cannot exceed the circumradius, that is

$$R \geq 2r.$$

This is an immediate consequence of Euler's identity $d^2 = R^2 - 2Rr$ (d is the distance of the two centres), but there are several proofs around. It is also related to various triangle inequalities and relations, for example

$$\cos\alpha + \cos\beta + \cos\gamma \leq \frac{3}{2},$$

$$\cos\alpha + \cos\beta + \cos\gamma = 1 + \frac{r}{R},$$

$$(-a+b+c)(a-b+c)(a+b-c) \leq abc.$$

The equality $R = 2r$ holds only in equilateral triangles.

[38] *Parallel chords.* Consider a circle about the origin on the Argand diagram and denote its four points by the complex numbers a, b, c and d. The chords connecting a to b and c to d are parallel if

$$ab = cd.$$

Assume that the counterclockwise order of the points is a, b, c and d. The corresponding chords are then parallel if and only if the arcs $\overset{\frown}{b,c}$ and $\overset{\frown}{d,a}$ are equal. Denote the central angle of these arcs by φ and let $e = \cos\varphi + i\sin\varphi$. Multiplication by e is hence a rotation by φ about the origin.

$$be = c \quad \text{and} \quad de = a,$$

from which

$$\frac{be}{de} = ca, \qquad \frac{b}{d} = \frac{c}{a}, \qquad \text{that is} \qquad ab = cd,$$

and the converse of the argument is also valid.

[39] *n-dimensional vectors.* Ordered n-tuples $(a_1, a_2, \ldots, a_n)$ of real numbers are sometimes called n-dimensional vectors and they are then denoted a single bold face letter: $\mathbf{a}(a_1, a_2, \ldots, a_n)$; the numbers a_i are then the coordinates of the vector. There is a natural way to perform algebraic operations between n-dimensional vectors as follows:

addition: the sum of the vectors $\mathbf{a}(a_1, a_2, \ldots, a_n)$ and $\mathbf{b}(b_1, b_2, \ldots, b_n)$ is

$$\mathbf{a} + \mathbf{b}(a_1 + b_1, a_2 + b_2, \ldots, a_n + b_n).$$

Subtraction: $\mathbf{a} - \mathbf{b}(a_1 - b_1, a_2 - b_2, \ldots, a_n - b_n).$

Multiplication by the real number λ: $\lambda\mathbf{a}(\lambda a_1, \lambda a_2, \ldots, \lambda a_n).$

Scalar, or dot product: $\mathbf{ab} = a_1 b_1 + a_2 b_2 + \ldots + a_n b_n.$

The fundamental algebraic laws are as follows:

$$\mathbf{a} + \mathbf{b} = \mathbf{b} + \mathbf{a}, \qquad \mathbf{ab} = \mathbf{ba}, \qquad \lambda\mathbf{a} = \mathbf{a}\lambda, \qquad \lambda(\mu\mathbf{a}) = (\lambda\mu)\mathbf{a} = \lambda\mu\mathbf{a},$$

$$(\lambda + \mu)\mathbf{a} = \lambda\mathbf{a} + \mu\mathbf{a}, \qquad \lambda(\mathbf{a} + \mathbf{b}) = \lambda\mathbf{a} + \lambda\mathbf{b}, \qquad \lambda(\mathbf{ab}) = (\lambda\mathbf{a})\mathbf{b} = \mathbf{a}(\lambda\mathbf{b}),$$

$$\mathbf{a}(\mathbf{b} + \mathbf{c}) = (\mathbf{b} + \mathbf{c})\mathbf{a} = \mathbf{ab} + \mathbf{ac} \qquad \text{(distributive law).}$$

Further notations and concepts: $\dfrac{\mathbf{a}}{\lambda} = \dfrac{1}{\lambda}\mathbf{a}$. The vector $\mathbf{0}(0, 0, \ldots, 0)$ is called zero vector; the product of two equal vectors is called the square of the given vector and abbreviated accordingly: $\mathbf{aa} = \mathbf{a}^2 = a_1^2 + a_2^2 + \ldots + a_n^2.$

The distributive law also holds if both factors have several terms; in particular:

$$(\mathbf{a}_1 + \mathbf{a}_2 + \ldots + \mathbf{a}_n)^2 = \mathbf{a}_1^2 + \mathbf{a}_2^2 + \ldots + \mathbf{a}_n^2 + 2(\mathbf{a}_1\mathbf{a}_2 + \mathbf{a}_1\mathbf{a}_3 + \ldots + \mathbf{a}_{n-1}\mathbf{a}_n).$$

The proof of any one of the above relations can be done by expanding them in terms of coordinates. For a further application see also [22].

[40] *Weighed means.* The notion of weighed means is a generalization of the notion of means.

Assign, as its weight, to each of the real numbers $a_1, a_2, \ldots, a_n$ a positive number s_i, respectively. The weighed arithmetic mean (or weighed average) of the numbers $a_1, a_2, \ldots, a_n$ is then

$$A_s = \frac{s_1 a_1 + s_2 a_2 + \ldots + s_n a_n}{s_1 + s_2 + \ldots + s_n};$$

their weighed geometric mean is

$$G_s = \sqrt[s_1 + s_2 + \ldots + s_n]{a_1^{s_1} a_2^{s_2} \ldots a_n^{s_n}};$$

the weighed harmonic mean is

$$H_s = \frac{s_1 + s_2 + \ldots + s_n}{\frac{s_1}{a_1} + \frac{s_2}{a_2} + \ldots + \frac{s_n}{a_n}},$$

and, finally, the weighed quadratic mean is

$$Q_s = i\frac{s_1 a_1^2 + s_2 a_2^2 + \ldots + s_n a_n^2}{s_1 + s_2 + \ldots + s_n}.$$

Togeteher these weighed means also obey the well known chain of inequalities between ordinary means, namely

$$H_s \leq G_s \leq A_s \leq Q_s.$$

The proof follows a general pattern. The first step it is straightforward: if the weights s_i are whole numbers then there is nothing new here, the weighed means can be conceived as ordinary means of appropriate number of copies of each number: there are s_i occurences of the number a_i. If the weights are rational then, as the following example shows, the issue can be reduced to the previous case. Let's see how to do this in the $A_s \geq G_s$ inequality for two terms; let the weights be $s_1 = \dfrac{p_1}{q_1}$ and $s_2 = \dfrac{p_2}{q_2}$ (p_i and q_i are positive integers). Then

$$A_s = \frac{\frac{p_1}{q_1}a_1 + \frac{p_2}{q_2}a_2}{\frac{p_1}{q_1} + \frac{p_2}{q_2}} = \frac{p_1 q_2 a_1 + p_2 q_1 a_2}{p_1 q_2 + p_2 q_1} \geq {}^{p_1 q_2 + p_2 q_1}\!\!\sqrt{a_1^{p_1 q_2} \cdot a_2^{p_2 q_1}} =$$

$$= {}^{\frac{p_1}{q_1} + \frac{p_2}{q_2}}\!\!\sqrt{a_1^{\frac{p_1}{q_1}} \cdot a_2^{\frac{p_2}{q_2}}} = G_s.$$

Finally, if there happen to be irrational numbers among the weights, then one should invoke standard continuity arguments. The point is that the means are continuous functions of the weights and irrational numbers can be approximated to arbitrary precision by rationals.

[41] *Trigonometric form of Ceva's theorem and an application.* If the lines a', b', c' are dividing the angles α, β, γ of the triangle ABC into the parts α_1 and α_2, β_1 and β_2, γ_1 and γ_2, respectively (*Figure 41.1*) then the lines a', b', c' are concurrent if and only if

(1) $$\frac{\sin \alpha_1 \sin \beta_1 \sin \gamma_1}{\sin \alpha_2 \sin \beta_2 \sin \gamma_2} = 1.$$

Assume first that the three lines in question are passing through a common point P. By the sine rule in the triangles ABP, BCP, CAP respectively

$$\frac{PA}{PB} = \frac{\sin \beta_1}{\sin \alpha_2}, \qquad \frac{PB}{PC} = \frac{\sin \gamma_1}{\sin \beta_2},$$

$$\frac{PC}{PA} = \frac{\sin \alpha_1}{\sin \gamma_2}.$$

and the product of the three equalities yields (1).

Assume, for the converse, that the lines a', b', c' divide the angles of the triangle according to (1). Denote the intersection of the lines a' and b' by P' and suppose that $P'C$ cuts the angle γ into the parts γ_1' and γ_2'. We have already seen that

$$\frac{\sin \alpha_1 \sin \beta_1 \sin \gamma_1'}{\sin \alpha_2 \sin \beta_2 \sin \gamma_2'} = 1,$$

which, when compared to (1), yields $\dfrac{\sin \gamma_1}{\sin \gamma_2} = \dfrac{\sin \gamma_1'}{\sin \gamma_2'}$, that is

$$\frac{\sin(\gamma - \gamma_2)}{\sin \gamma_2} = \frac{\sin(\gamma - \gamma_2')}{\sin \gamma_2'}, \qquad \text{or}$$

$$\sin \gamma_2'(\sin \gamma \cos \gamma_2 - \cos \gamma \sin \gamma_2) = \sin \gamma_2(\sin \gamma \cos \gamma_2' - \cos \gamma \sin \gamma_2'),$$

$$\sin \gamma_2' \cos \gamma_2 = \sin \gamma_2 \cos \gamma_2',$$

$$\sin(\gamma_2' - \gamma_2) = 0.$$

This implies $\gamma_2 = \gamma_2'$ and $\gamma_1 = \gamma_1'$ and thus P' and P are identical, the proof is complete.

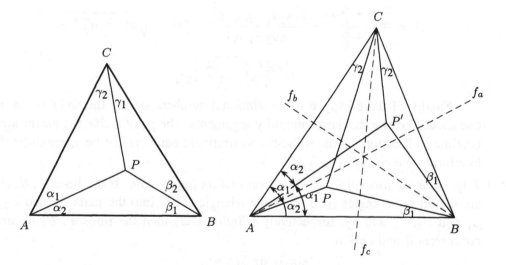

Figure 41.1.   Figure 41.2.

An immediate consequence is the theorem used in the solution of Qu. No 2. in 1996: if P is an interior point of the triangle ABC and the line PA is reflected in the bisector of $A\angle$, PB is reflected to the bisector of $B\angle$, finally PC is reflected in the bisector of $C\angle$, then the reflected lines are concurrent. Indeed, the parts α_1 and α_2, β_1 and β_2, γ_1 and γ_2 are swapped under the reflections and hence (1) remains valid, the mirror lines are also passing through a common point (*Figure 41.2.*).

This assertion can be proved in a more general form without the Ceva-trigonometry machinery using reflections only; if the lines through the vertices of a triangle belong to a pencil (i.e. they are concurrent or parallel) then the same holds for the mirror images in the corresponding angle bisectors.

[42] *An extension of the Erdős–Mordell inequality.* Let P, Q and S be interior points on the sides AB, BC, CA of the triangle ABC, respectively. Denote the intersection of the perpendiculars to AB at P and to BC at Q by Y and similarly, the intersection of the perpendiculars to BC at Q and to CA at S by Z and, finally, the intersection of the perpendiculars to CA at Q and to AB at P by X. If, additionally, the points X, Y and Z are interior to the triangle ABC then

$$AX + BY + CZ \geq XP + YP + YQ + ZQ + ZS + XS,$$

and equality holds if and only if the triangle ABC is equilateral and the each of the points X, Y, Z are at the centre of ABC. (*Figure 42.1.*).

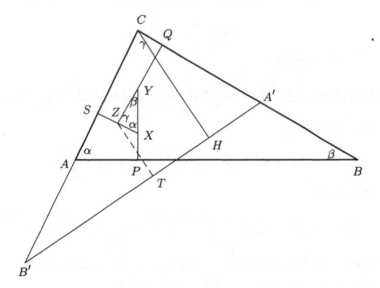

Figure 42.1.

The proof is following the demonstration in [33] of the original theorem. Denote by A', B' the mirror image of the vertices A and B in the interior bisector of the angle C, respectively; the feet of the perpendiculars to $A'B'$ from C and Z be H and T, respectively. The area of the triangle $A'B'C$ can be written in two different ways:

$$2a_{A'B'C} = 2a_{A'B'Z} + 2a_{CA'Z} + 2a_{B'CZ}.$$

With the compulsory notations $AB = A'B' = c$, $BC = B'C = a$, $CA = CA' = b$ this can be put as

(1) $$c \cdot CH = c \cdot ZT + b \cdot ZQ + a \cdot ZS;$$

here the length of ZT is negative if the line $A'B'$ separates the points Z and C. In any case we have the following inequality:

(∗) $$CZ + ZT \geq CH.$$

Hence

$$c \cdot CZ + c \cdot ZT \geq c \cdot CH,$$

$$c \cdot CZ \geq c \cdot CH - c \cdot ZT,$$

which, when combined with (1), implies

$$c \cdot CZ \geq b \cdot ZQ + a \cdot ZS,$$

$$CZ \geq \frac{b}{c} \cdot ZQ + \frac{a}{c} ZS.$$

Similarly

$$AX \geq \frac{c}{a} XS + \frac{b}{a} XP,$$

$$BY \geq \frac{a}{b} YP + \frac{c}{b} YQ.$$

Adding these inequalities

$$(2) \quad AX + BY + CZ \geq \left(\frac{a}{b} YP + \frac{b}{a} XP \right) + \left(\frac{b}{c} ZQ + \frac{c}{b} YQ \right) + \left(\frac{c}{a} XS + \frac{a}{c} ZS \right).$$

Apply now the identity

$$kK + nN = (k + n)\frac{K + N}{2} + (k - n)\frac{K - N}{2}$$

for the expression

$$\frac{a}{b} YP + \frac{b}{a} XP = \left(\frac{a}{b} + \frac{b}{a} \right) \frac{YP + XP}{2} + \left(\frac{a}{b} - \frac{b}{a} \right) \frac{YP - XP}{2}.$$

Observe that their angles being pairwise equal the triangles ABC and XYZ are similar. If λ is the scale factor of similarity then

$$\frac{YZ}{a} = \frac{ZX}{b} = \frac{XY}{c} = \lambda,$$

and also $YP - XP = XY = \lambda c$. By

$$(\ast\ast) \qquad \frac{a}{b} + \frac{b}{a} \geq 2.$$

we now get

$$\frac{a}{b} \cdot YP + \frac{b}{a} \cdot XP \geq YP + XP + \lambda \left(\frac{ca}{2b} - \frac{bc}{2a} \right).$$

Similarly

$$\frac{b}{c} \cdot ZQ + \frac{c}{b} YQ \geq ZQ + YQ + \lambda \left(\frac{ab}{2c} - \frac{ca}{2b} \right),$$

$$\frac{c}{a} \cdot XS + \frac{a}{c} \cdot ZS \geq XS + ZS + \lambda \left(\frac{bc}{2a} - \frac{ab}{2c} \right).$$

By (2) the sum of these inequalities implies the claim.

$$AX + BY + CZ \geq XP + YP + YQ + ZQ + ZS + XS.$$

If there is equality then by (**) $a = b = c$ and by (*) X, Y and Z are laying on the respective altitudes; summarizing the conditions of equality ABC has to be equilateral and each of X, Y, Z must be at its centre.

[43] *A property of equilateral triangles.* Let P be an arbitrary point on the arc AB not containing C of the circumcircle of the equilateral triangle ABC. Then $AP + BP = PC$.

Rotate, about A, the triangle APB by $60°$. If the image is ACP' then P' is on the segment CP because $ACP\angle = ABP\angle$ (inscribed angles intercepting the same arc AP). By the rotation the triangle APP' is equilateral and thus $AP = PP'$; since, on the other hand, $BP = CP'$ we get

$$PC = PP' + CP' = AP + BP,$$

indeed. (*Figure 43.1.*).

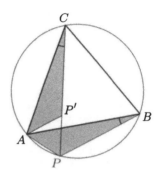

Figure 43.1.

We note that the claim is an immediate consequence of Ptolemy's theorem [24]; when applied to the cyclic quadrilateral $APBC$ it yields

$$AP \cdot BC + BP \cdot CA = AB \cdot PC,$$

and dividing through by the length of the side of the triangle we get the desired result.

[44] *The number of divisors.* The number of (positive) divisors of a positive integer n is denoted by $d(n)$. (One can also come across to the notation $\tau(n)$.) If the prime factorization of n is

$$n = p_1^{\alpha_1} \cdot p_2^{\alpha_2} \cdot \ldots \cdot p_r^{\alpha_r},$$

then every divisor of n is equal to

$$p_1^{\beta_1} p_2^{\beta_2} \cdots p_r^{\beta_r}$$

where $0 \leq \beta_i \leq \alpha_i$, and, conversely, for any such choice of the numbers β_i there is a divisor of the above form. Therefore, there are $\alpha_i + 1$ ways to set the value of β_i and, accordingly

$$d(n) = (\alpha_1 + 1)(\alpha_2 + 1) \ldots (\alpha_r + 1)$$

divisors of n, altogether. This shows that $d(n)$ depends on the list of indices only, not on the actual prime factors. As a consequence we note here that if a and b are coprime then

$$d(ab) = d(a)d(b),$$

the function d is multiplicative. This property is, of course, true for products of finitely many coprime factors.

[45] *Turán's graph theorem.* Paul Tur n proved the following theorem in 1941: let $n = q(k-1) + r$, where q, k, r are whole numbers such that $0 \le r < k - 1$. If there are more than

$$E = \frac{k-2}{2(k-1)}(n^2 - r^2) + \binom{r}{2},$$

edges in a simple graph of n vertices then the graph contains a complete subgraph of k vertices. The result is sharp, since for every n there exists a simple graph of n vertices and E edges with no complete subgraph of k vertices.